Physics for Applied Biologists

N. C. Hilyard and H. C. Biggin

Physics For Applied Biologists

N. C. Hilyard and H. C. Biggin

Department of Applied Physics
Sheffield City Polytechnic

University Park Press

First published 1977
by Edward Arnold (Publishers) Limited
London

First published in the USA by
University Park Press
233 East Redwood Street
Baltimore, Maryland

Library of Congress Cataloging in Publication Data

Hilyard, N. C.
Physics for applied biologists.

Includes index.
1. Physics. 2. Biological physics. I. Biggin,
H. C., joint author. II. Title.
QC21.2.H54   530   77-14979
ISBN 0-8391-1189-4

Photoset and printed in Malta by Interprint (Malta) Ltd.

Preface

The study of biological sciences is becoming increasingly quantitative in both theoretical and experimental aspects and there is need for biology students to have some knowledge of mathematics and the physical sciences in order to understand the behaviour and function of living matter. This text, written for both graduate and undergraduate students, is based on our experience of five years teaching of applied biologists and attempts to put physics in the biological context, and so provide the student with a working knowledge of the physical nature of his subject.

The earlier chapters assume a knowledge of physics and mathematics attainable at an advanced high school or introductory undergraduate level. New concepts are quickly introduced to maintain the interest of students with additional background knowledge. Our experience of students with no formal education in physics is that there are no insuperable barriers to adequate understanding of the content of the text. Much emphasis is placed on graphical interpretation, a technique used to overcome the necessary omission of differential and integral calculus. There are a liberal number of examples in each chapter, some of which include textual material to avoid excessive repetition. Also a selection of problems is given at the end of each chapter which should be treated as an integral part of the chapter content. They are designed to reinforce concepts introduced in the text by indicating how such concepts are applied in situations which the students will encounter. Answers to the numerical parts of each problem are given.

In this text physics is treated as an integral part of quantitative biology, equal emphasis being given to the measurement of physical quantities, the manipulation of quantitative information and the explanation of physical behavior at the molecular level. The International System of Units is used throughout. Certain symbols are used to signify more than one quantity. For example, E represents energy, electric field intensity and tensile modulus and μ denotes both reduced mass and linear absorption coefficient. Since these are accepted symbols in the various subject areas, we have used the conventional symbol rather than a unique but artificial one. The significance of all symbols is clearly indicated as they appear in the text.

Sheffield   NCH
1976   HCB

Contents

1

Force, Motion and Energy

Living organisms and their environment are in a state of continuous motion. Even the small particles which comprise an inanimate object, such as a piece of wood, are moving. This motion is produced by forces, such as the gravitational force which causes rain-drops to fall from the sky or the electrostatic force which binds atoms and molecules together. If a body or particle is moving or is situated in a force field it possesses energy. Energy can take many different forms, e.g. mechanical, thermal and chemical. The transmission and the transformation of energy into its different forms is of fundamental importance in the function of a biological system.

1.1 Position, velocity and acceleration

In order to describe the state of a system, such as its location in space, its movement or its temperature, it is necessary to define the quantities used to describe this state and the frame of reference in which these quantities apply. Both the units and the frame of reference are arbitrary. For example we may say that a road junction is three miles north-east of a city centre. We have defined the distance unit as miles, the frame of references as two mutually perpendicular directions, north and east, and the origin of the frame of reference at the centre of the city. Alternatively we may say that a tree is a number of metres from the boundary hedges of a garden. In this case the unit of distance is the metre and the frame of reference is the boundary hedges. Another example is the measurement of temperature using Celsius (centigrade) and Fahrenheit thermometers. Although they may measure the temperature of the same body, the quantities that these thermometers indicate are different because the origin (i.e. zeros) of the scales and the basic units (degrees Celsius or degrees Fahrenheit) are different.

Various systems of units have been devised in order to communicate quantitative information. The one used in this book is called the SI system in which the basic quantity of length is the metre, the quantity of mass is the kilogram and the quantity of time is the second. These quantities are abbreviated as m, kg and s respectively. Unfortunately other systems of units are still in use. The SI units, their abbreviations and their relationship with other system units are given in Appendix A together with the most commonly used multiples and submultiples of the units. Whenever quantitative information is manipulated to work out the results of an experiment or to solve a numerical problem, all quantities should be converted to SI units before performing calculations.

Throughout the book you will find quantities expressed in non-SI form. This is to familiarize you with the alternative units and their conversion into the SI system.

Some of the ways in which the location of a particle P can be described are shown in Fig. 1.1. In the simplest case, Fig. 1.1a, P can be defined by a single quantity x in a one-dimensional frame of reference (the x-axis) which has origin O. The distance OP is equal to the position coordinate x. In more complex situations it is necessary to use two (x, y)- or three (x, y, z)-dimensional coordinate systems. The position coordinates of P are (x, y) and (x, y, z) respectively and the corresponding distances OP are $\sqrt{x^2 + y^2}$ and $\sqrt{x^2 + y^2 + z^2}$.

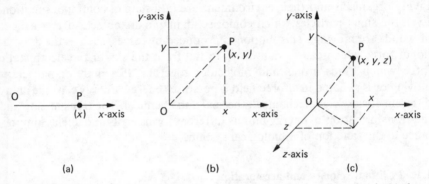

Figure 1.1 Three coordinate systems: (a) one-dimensional; (b) two-dimensional; (c) three-dimensional.

In all natural systems the particles are in continuous motion. In order to describe this motion it is necessary to introduce another dimension; time. It is difficult to work in four-dimensional coordinates. Fortunately it is possible to consider separately the motion in the x, y and z directions, so that the four-dimensional system can be reduced to three two-dimensional systems $(x - t, y - t, z - t)$. An example of this is the jump of the flea. We may define a three-dimensional coordinate system, Fig. 1.1c, with origin O at the initial location of the flea. When the flea jumps its motion can be recorded using a high-speed cine-camera and its position coordinates x, y and z determined at any time. The three components of the motion can be considered separately. The vertical (y) position of the flea at different times after the initiation of the jump is shown in Fig. 1.2a. By measuring the distance between the flea and the ground (defined as the origin of the y coordinate), the variation of y with t can be determined and plotted on a y–t graph. This is shown in Fig. 1.2b. A quantity, such as the elevation y, consists of a number multiplied by a unit. The label $y/10^{-3}$ m on the position axis gives the measured quantity; e.g. a coordinate of $1 \cdot 0$ on this axis indicates the elevation is $1 \cdot 0 \times 10^{-3}$ m. If sufficient measurements are taken (these are called data points) a smooth curve can be drawn which shows the position of the flea at any time during the time period of the measurement. This curve indicates that the altitude of the flea increases slowly over the first millisecond ($1 \cdot 0$ ms $= 1 \cdot 0 \times 10^{-3}$ s). After this its vertical position increases in proportion to time.

If at some time t_0 the flea is in position y_0 and at some time t later it is at

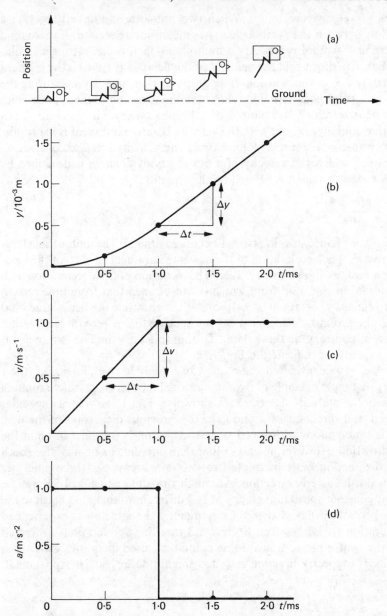

Figure 1.2 (a) The jump of the flea in the vertical direction. The variation of (b) position, (c) velocity and (d) acceleration with time. (Adapted with permission from Rothschild M., Schlien Y., Parker K., Neville C. and Sternberg S. *Sci. Am.*, **229**, No. 5, 1973.)

y it has travelled a distance $\Delta y = y - y_0$ (in the y direction) in a time $\Delta t = t - t_0$. The symbol Δ signifies a small change in a quantity, Δy being a change in position and Δt a change in time. The velocity during this part of the motion is defined as

$$v = \frac{\text{distance travelled}}{\text{time taken}} = \frac{\Delta y(\text{m})}{\Delta t(\text{s})} \qquad (1.1)$$

The units of velocity are m s^{-1}. When two units are multiplied together a space is left between the symbols, e.g. m s means metre second. If there is no space, the first symbol represents a multiplying factor, e.g. ms means millisecond. Equ. 1.1 shows that the velocity of the flea at any time t is the gradient (or slope), $\Delta y/\Delta t$, of the position–time graph at that time. By measuring the gradient at several positions along the curve the variation of the velocity with time can be determined and plotted on a graph, as shown in Fig. 1.2c. It can be seen that initially, i.e. at $t = 0$, the velocity is zero but increases in proportion to t up to $t = 1 \cdot 0$ ms. At longer times the velocity remains constant.

The rate at which the velocity of a moving body changes is described by the acceleration, a, which is defined by the equation

$$a = \frac{\Delta v (\text{m s}^{-1})}{\Delta t (\text{s})} \qquad (1.2)$$

where Δv is the change in velocity that occurs in time Δt. The units of acceleration are m s^{-2}. The acceleration of the flea is the gradient, $\Delta v/\Delta t$, of the velocity–time graph as shown in Fig. 1.2c. The variation of the acceleration with time is plotted in Fig. 1.2d from which it can be seen that from time zero to $1 \cdot 0$ ms a remains constant at 1×10^3m s^{-2}. At this time the flea's large hind legs leave the ground. At longer times the acceleration is zero (at least within the limits of accuracy of these data). During this time the flea is moving at constant velocity, as indicated in Fig. 1.2c.

So far we have considered the motion of a body in one dimension (the y direction). A simple example of two-dimensional motion is an athlete running around a horizontal circular track as shown in Fig. 1.3a. Velocity specifies both speed and direction of motion. In the previous discussion of the flea, velocity is synonymous with speed since the motion is specified to be in the vertical direction. However, in cases other than one-dimensional motion, both speed and direction must be quoted to define velocity, e.g. three miles per hour in a north-easterly direction. Although the athlete in Fig. 1.3a may be moving at constant speed his velocity at P is different to that at Q because the direction of motion has changed. Consequently the athlete has experienced an acceleration. It is shown in Section 1.3 that this acceleration is directed towards the centre of the track. Some commonly used quantities with which it is necessary to specify direction as well as magnitude are listed in Appendix B.

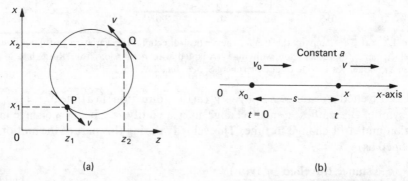

(a)   (b)

Figure 1.3 (a) Motion in two dimensions. (b) Motion in one dimension with constant acceleration.

For a body moving in one dimension with constant acceleration it is possible to set up several equations of motion which relate position, velocity, time and acceleration. Consider for example motion in the x direction at constant acceleration, a, as shown in Fig. 1.3b. The position coordinate and the velocity at time $t = 0$ are x_0 and v_0. From the definition of acceleration, Equ. 1.2, $\Delta v = v - v_0 = at$ so that the velocity at time t is $v = v_0 + at$. The other equations of motion can be established in the same way. The most useful forms are

$$v = v_0 + at \tag{1.3}$$

$$x = x_0 + (v + v_0)t/2 \tag{1.4}$$

$$x = x_0 + v_0t + at^2/2 \tag{1.5}$$

$$v^2 = v_0^2 + 2a(x - x_0) \tag{1.6}$$

Remember that the distance travelled $s = \Delta x = x - x_0$.

Example 1.1

During the first 1 ms of its jump motion the flea of Fig. 1.2 experiences a constant acceleration of 1×10^3m s^{-2}. What is its position and velocity when $t = 1$ ms?

From Equ. 1.5, $y = y_0 + v_0t + at^2/2$, with $y_0 = 0$, $v_0 = 0$ and $a = 1 \times 10^3$ m s^{-2}.

$$y = 1 \times 10^3 \text{ m s}^{-2} \times (1 \times 10^{-3} \text{ s})^2/2$$

$$y = 5 \times 10^{-4} \text{ m} = 0.5 \text{ mm}$$

Whenever quantities are introduced into an equation the units should be stated and manipulated in the usual way in order to obtain the units of the calculated quantity. If these units are not the same as the accepted units for this quantity the calculations should be checked.

From Equ. 1.3 the velocity v in the y direction is

$$v = 1 \times 10^3 \text{ m s}^{-2} \times 1 \times 10^{-3} \text{ s}$$

$$v = 1 \text{ m s}^{-1}$$

1.2 Forces producing acceleration

Accelerations which occur during motion are produced by forces. It is a common experience that when a stationary object is pushed sufficiently hard it moves. A force F has been applied, the object has changed its velocity and has consequently experienced an acceleration. The acceleration is proportional to the applied force, $F \propto a$, which can be written as

$$F = Ma \tag{1.7}$$

The constant of proportionality M is called the mass. Since $a = F/M$, the more massive the object the larger the force required to produce a given acceleration. Equ. 1.7 is called Newton's second law of motion. The unit of force is the newton (N) which is the force necessary to produce an acceleration of 1 m s^{-2} in an object of mass 1 kg.

Example 1.2

A passenger of mass 70 kg in a motor vehicle initially travelling at 100 kilo-metres per hour is attached to the frame of the vehicle by a belt that can withstand a force of $1 \cdot 0 \times 10^4$ N without breaking. The vehicle hits an immov-able object and stops in $1 \cdot 0$ s, Assuming that the acceleration during impact is constant what is the motion of the passenger?

The initial speed of the passenger in the vehicle is

$$v = \frac{100 \times 10^3 \text{ m h}^{-1}}{60 \times 60 \text{ s h}^{-1}} = 28 \text{ m s}^{-1}$$

From Equ. 1.3 with $v = 0$, $v_0 = 28 \text{ m s}^{-1}$ and $t = 1 \cdot 0$ s, the acceleration during impact is $a = -v_0/t$ so that

$$a = \frac{-28 \text{ m s}^{-1}}{1 \cdot 0 \text{ s}} = -28 \text{ m s}^{-2}$$

The minus sign indicates that the acceleration is in a direction opposite to the direction of movement. From Equ. 1.7 the magnitude of the force exper-ienced by the belt is

$$|F| = 70 \text{ kg} \times 28 \text{ m s}^{-2} = 2 \cdot 0 \times 10^4 \text{ N}$$

Consequently the belt will break and the passenger will be thrown forwards.

Consider an object suspended in space with two equal and opposite forces F applied to it as shown in Fig. 1.4a. (Remember that a force has direction as well as magnitude.) The magnitude of the force F_R acting to the right is equal to the magnitude of the force F_L acting to the left so that the net force experienced by the object is zero. Since $F_R = F$ and $F_L = -F$ this can be written as $F_R + F_L = F - F = 0$. Since the net force is zero the body does not accelerate, Equ. 1.7, and it is said to be in a state of equilibrium. This is Newton's first law of motion. When a system is in equilibrium its state does not change. The state of a system can refer to the motion of a body or its shape or its temperature or even the rate of a chemical reaction.

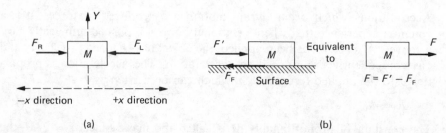

Figure 1.4   (a) The equilibrium of forces. (b) An unbalanced force.

If a body rests on a surface, Fig. 1.4b, there is an interaction between the body and the surface which generates a force, F_F, called the frictional force. The frictional forces always oppose the motion or the intended motion. When a force F' is applied to the body the net, or unbalanced, force experienced by the body is $F = F' - F_F$. It is this unbalanced force which upsets the state of equilibrium and causes motion with acceleration $a = F/M$.

The product of the velocity of a body and its mass is called the linear momentum p,

$$p = Mv \tag{1.8}$$

Since velocity has direction the linear momentum has direction. (Although the same symbol, v, is used to represent velocity and speed it is necessary to examine each particular situation to determine if the direction of motion is important.) If no unbalanced forces act on the body, $a = 0$ (from $a = F/M$) and the velocity is constant. If its mass remains unchanged, the body will have constant linear momentum. The concept of momentum is important in describing the collision of two or more bodies, the total momentum of the system being the same after collision as before if no unbalanced force is acting. This is the principle of the conservation of momentum. Consider a system made up of two bodies one of mass M_1 moving with speed u_1 and the other of mass M_2 with speed u_2 as shown in Fig. 1.5a. Since no external forces act on this system momentum is conserved when these bodies collide and the total momentum before collision is equal to the total momentum after collision:

$$M_1u_1 + M_2u_2 = M_1v_1 + M_2v_2 \tag{1.9}$$

where v_1 and v_2 are the speeds of M_1 and M_2 after the collision. In the particular collision shown in Fig. 1.5b where the second body is initially at rest ($u_2 = 0$) and the two bodies move off together after collision ($v_2 = v_1$), Equ. 1.9 reduces to $M_1u_1 = (M_1 + M_2)v_1$ or

$$v_1 = M_1u_1/(M_1 + M_2)$$

For motion in more than one direction the conservation of momentum is applied to each component, i.e. the momentum in the x, y or z directions.

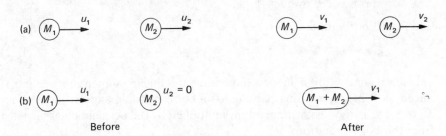

Figure 1.5 One-dimensional collisions with conservation of momentum.

1.3 Motion in a circle

A common case of two-dimensional motion is when a particle (mass M) moves in a circular path (radius r) at uniform speed (v). This is called uniform circular motion and is shown in Fig. 1.6a. At some time the particle is at position P and Δt seconds later it has moved a distance Δs around the circumference to Q. The time required for this is $\Delta t = \Delta s/v$ so that the time to complete one revolution, a distance $2\pi r$, is

$$T = 2\pi r/v$$

T is the period of the motion. The number of revolutions completed each second is the frequency f of rotation:

$$f = 1/T = v/2\pi r \tag{1.10}$$

which has units s^{-1} or hertz (Hz).

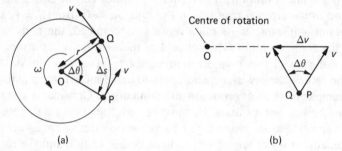

Figure 1.6   (a) Uniform circular motion. (b) Velocity triangle.

Just as the linear velocity v is the rate of change of position with time so the angular velocity ω (sometimes called the angular frequency) is defined as the rate of change of angle with time:

$$\omega = \Delta\theta/\Delta t \tag{1.11}$$

The unit of angle is the radian, an angle $\Delta\theta$, Fig. 1.6a, being equal to $\Delta s/r$ radians. For a complete circle ($\Delta s = 2\pi r$), there are 2π radians so that one radian is equal to 360 degrees/2π which is 57·3 degrees. Since one revolution takes T seconds we have from Equ. 1.11

$$\omega = 2\pi/T = 2\pi f \,(\text{rad s}^{-1})$$

and from Equ. 1.10, $\omega = 2\pi v/2\pi r = v/r$ so that

$$v = \omega r \tag{1.12}$$

In discussing motion in one dimension the linear momentum was introduced, $p = Mv$. A rotating particle has a corresponding angular momentum L, where L is the linear momentum multiplied by the perpendicular distance from the axis of rotation:

$$L = Mvr = M\omega r^2 \tag{1.13}$$

Rotation is caused by a torque, which is the product of the applied force and the perpendicular distance between the point of application of the force and the axis of rotation. Angular momentum is conserved when there are no unbalanced torques.

As explained in Section 1.1, although the speed of a particle in uniform circular motion is constant, the velocity is continuously changing because the direction of motion changes. The velocity at P can be represented by an arrow drawn perpendicular to the radius r (tangential to the circumference). The length of this arrow represents the speed v and its direction is the direction of the velocity. A similar arrow can be drawn for the point Q. When the bases of these arrows are put together, the angle between them being $\Delta\theta$, and their tips joined a velocity triangle is formed. The line Δv represents the size and the direction of the change in the velocity as the particle moves from P to Q. It can be seen that the change in velocity is directed towards the centre of the circle O, so the acceleration, $\Delta v/\Delta t$, experienced by the particle is directed towards the centre of the circle. For small values of $\Delta\theta$, it can be seen from Fig. 1.6b that, $\Delta v/v = \Delta\theta$ and since $\Delta\theta = \Delta s/r$ we have $\Delta v/v = \Delta s/r$ or $\Delta v = \Delta s(v/r)$. Dividing both sides by Δt gives $(\Delta v/\Delta t) = (\Delta s/\Delta t)(v/r)$. From the definitions of a and v and Equ. 1.12, $a = v(v/r)$ or

$$a = v^2/r = \omega^2 r \tag{1.14}$$

This acceleration, which is directed towards the centre of rotation, is called the centripetal acceleration and is provided by the centripetal force $F_c = Ma$. From Equ. 1.14

$$F_c = Ma = Mv^2/r \tag{1.15}$$

which increases as the mass and speed increase and as the radius of the motion decreases. A hammer-thrower rotates a heavy mass in a circular path. The centripetal force is exerted by the thrower and is transmitted to the mass by the wire joining the mass and the thrower's hands. When the thrower releases the wire this centripetal force disappears and the mass moves in a straight line which is tangential to the initial circular motion, see the velocity arrows in Fig. 1.6a.

Example 1.3

A hammer-thrower rotates a 7·3 kg shot on a wire of length 1·2 m at 2·0 revolutions per second. If the length of his arms is 0·80 m, calculate (a) the period of rotation; (b) the linear speed of the shot; (c) the angular momentum of the shot; and (d) the tension in the wire.

(a) With an angular velocity of 2·0 revolutions per second, the time for one revolution $T = 0·50$ s.

(b) Using Equ. 1.12 where $\omega = 2\pi \text{rad}/T(s) = 4·0\pi$ rad s^{-1} and $r = (1·2 \text{ m} + 0·8 \text{ m}) = 2·0$ m gives

$$v = 4·0\pi \text{ s}^{-1} \times 2·0 \text{ m} = 25 \text{ m s}^{-1}$$

(c) Substituting for $M(= 7·3$ kg$)$, v and r into Equ. 1.13 we have

$$L = 7·3 \text{ kg} \times 25 \text{ m s}^{-1} \times 2·0 \text{ m} = 3·7 \times 10^2 \text{ kg m}^2 \text{ s}^{-1}$$

(d) The tension in the wire provides the centripetal force F_c which constrains the shot to its circular path. Using Equ. 1.15, the tension in the wire is

$$F_c = \frac{7 \cdot 3 \text{ kg} \times (25 \text{ m s}^{-1})^2}{2 \cdot 0 \text{ m}} = 2 \cdot 3 \times 10^3 \text{ N}$$

The centripetal force is the basis of the centrifuge, an instrument widely used for separating particles of different mass suspended in a liquid. Fig. 1.7a shows the principle of centrifuge operation. A sample of the liquid is placed in a tube which is rotated at high speed. The tube and rotor arm are connected to the centre of rotation and experience a centripetal acceleration. The particles are not so connected but are constrained to move in a circular path by the liquid and the walls of the tube. The liquid is not able to trans-mit the steady centripetal force (see § 10.5) needed to maintain the particles in the circular path and they move outwards, away from the centre of rota-tion. This motion is impeded by viscous drag which depends upon the visco-sity of the liquid and the size of the particle. The result is that the high-mass particles move quickly to the bottom of the tube to form a separated pellet. Fig. 1.7b shows schematically the variation in the distribution of particles with time for an initially homogeneous suspension of particles of different masses. Very high angular velocities (several thousand revolutions per second) are achieved in the ultracentrifuge. The corresponding out-of-balance forces are sufficiently large to separate mixtures of molecules of different molecu-lar weights. This technique is used to measure the molecular weight of syn-thetic and biopolymers.

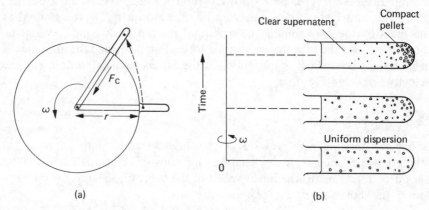

Figure 1.7 The principle of operation of the centrifuge.

1.4 Force fields

There are several fundamental forces that occur in nature. Most natural phe-nomena can be explained in terms of these forces. The forces of importance in our studies are the gravitational, the electrostatic and the magnetic forces.

When a person jumps from the surface of the earth he quickly returns. Initially he is moving in the upwards direction but a short time later he is moving downwards. Consequently he has experienced an acceleration, direc-ted towards the earth, so he must have been under the influence of some external force (from $F = Ma$). This force, between the person and the earth, is the gravitational force that exists between all masses. The force of attraction between two masses, M_1 and M_2, separated by a distance r is

$$F_G = GM_1M_2/r^2 \tag{1.16}$$

where G is the universal constant of gravitation. It is not necessary for the two bodies, M_1 and M_2, to be in contact for interaction to occur. Action can take place at a distance. This can be explained using the concept of a force field. The mass M_1 creates a force field which radiates outwards from the mass and pervades all space. The intensity of this field $I = GM_1/r^2$ decreases with increasing value of r (this is known as an inverse square law). When a mass M_2 is situated in a gravitational field of intensity I it experiences an attractive force $F_G = IM_2$ along the line joining the centres of the two masses.

The gravitational force on a body of mass M situated on the surface of the earth (mass M_e, radius R) is, from Equ. 1.16, $F_G = M(GM_e/R^2)$. Comparing this equation with $F = Ma$ it can be seen that the mass experiences an acceleration $a = GM_e/R^2$ directed towards the centre of the earth. This is called the acceleration due to gravity, $g = GM_e/R^2$. There are small variations in the value of g according to the location on the earth because of the ellipsoidal shape of the earth and local variations in the elevation and the composition. The standard value of g is 9·81 m s^{-2}. This gravitational force, $F = Mg$, is called the weight of the body, the larger the mass the larger the weight. The average man has a mass of 70 kg and his weight in the gravitational field of the earth is

$$F_G = Mg = 70 \text{ kg} \times 9\cdot81 \text{ m s}^{-2} = 6\cdot9 \times 10^2 \text{ N}$$

Although the gravitational force on every body on earth acts downwards we are not pulled underground because there is a reaction force due to the earth's surface acting upwards. This reaction force, which is shown in Fig. 1.8a, is communicated to the centre of gravity by the legs and the hips. Newton's third law of motion states that the action of a force and its reaction are equal and opposite. The reaction force is equal to the weight so that there is a balance of forces and the body is in equilibrium.

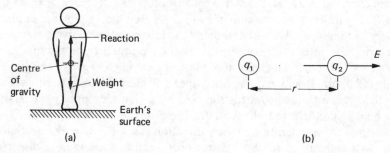

Centre of gravity — Reaction — Weight — Earth's surface

(a)   (b)

Figure 1.8 (a) Action and reaction are equal and opposite at the surface of the earth. (b) The interaction between two charged particles.

Example 1.4

The flea of Fig. 1.2 leaves the surface of the earth with an upwards velocity (the $+y$ direction) of 1·0 m s^{-1}. What is the maximum height H reached by the flea?

After leaving the surface the force experienced by the flea is the gravitational force which produces an acceleration g in the downwards, $-y$, direction. This force slows down the flea until it eventually stops, at the position of maximum height, and it then falls back to earth with acceleration g. From Equ. 1.6 $v^2 = v_0^2 + 2a(y - y_0)$. Substituting $v = 0$, $v_0 = 1 \cdot 0$ m s^{-1}, $y_0 = 0$ and $a = -g$ (the minus sign is because the acceleration is in the $-y$ direction) we have $H = -v_0^2/(-2g) = v_0^2/2g$

$$H = \frac{(1 \cdot 0 \text{ m s}^{-1})^2}{2 \times 9 \cdot 81 \text{ m s}^{-2}} = 5 \cdot 1 \times 10^{-2} \text{ m} = 51 \text{ mm}$$

Jump heights of up to 90 mm have been recorded.

Materials are made up of electrically charged particles. Electrical charge is a fundamental physical quantity rather like mass. There are two types of charge; positive and negative. The unit of charge is the coulomb, which has the symbol C. When two particles carrying charges q_1 and q_2 are separated by a distance r, Fig. 1.8b, there is a force between these particles given by the equation

$$F_E = k \, q_1 q_2/r^2 \tag{1.17}$$

which acts along the line joining the centres of the particles. k is a constant of the medium separating the charges. For vacuum or air $k = 9 \times 10^9$ N m^2 C^{-2}. Equ. 1.17 is called Coulomb's law. If the charges are of the same sign, i.e. both positive or both negative, the force is repulsive and if they are of opposite sign the force is attractive. This electrostatic interaction occurs because of the electrostatic field of intensity $E = k \, q_1/r^2$ due to charge q_1. Charge q_2 interacts with this field and it experiences a force F_E given by

$$F_E = Eq_2 \tag{1.18}$$

Thus the units of electric field intensity E are N C^{-1}. If the charges are free to move they will experience an acceleration $a = F_E/M = qE/M$ because of this force field.

An electric field E is established when a potential difference V is set up between two parallel plates as shown in Fig. 1.9 (see § 5.1). The upper plate effectively becomes positively charged and the lower plate negatively charged. The electric field intensity between the plates separated by a distance d is $E = V/d$ (N C^{-1}). When an electron of mass M carrying a negative charge $-e$ coulomb (Chapter 3) and moving with speed v_z along the z-axis enters this electric field it is attracted towards the positively charged upper plate. If the electric field is uniform the electrostatic force in the $+y$ direction is $F_E = eE = eV/d$ and the acceleration in the y direction is $a = F_E/M = eV/Md$. Since there is no acceleration in the z direction (because there is no z-directed force) the electron moves with constant speed v_z in this direction and the time spent in the electric field is $t = L/v_z$, where L is the length of the plates. Since the initial y position $y_0 = 0$ and the initial velocity in the y direction $v_{0y} = 0$ the y position as the electron leaves the plates is, from Equ. 1.5, $y = at^2/2$. Substituting for a and t gives

$$y = eVL^2/2Mdv_z^2 \tag{1.19}$$

The displacement y of the electron is directly proportional to the potential difference V between the plates. When the electron is outside the electric field it does not experience a force so that it travels in a straight line with constant velocity. The deviation of a beam of electrons by an electric field is employed in the cathode ray oscilloscope (§ 6.5).

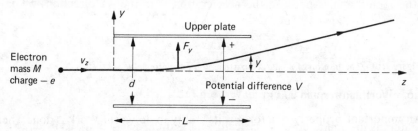

Figure 1.9   The motion of an electron in an electric field.

Example 1.5

An electron of mass $9 \cdot 1 \times 10^{-31}$ kg moving at $1 \cdot 0 \times 10^{6}$ m s^{-1} in the horizontal direction enters the region between two horizontal parallel plates $0 \cdot 020$ m apart. If the potential difference between the plates is $5 \cdot 0$ V and the plates are $1 \cdot 0 \times 10^{-2}$ m long, find the vertical distance travelled by the electron in the region of the field.

Using Equ. 1.19 where $V/d = 5 \cdot 0$ V$/2 \cdot 0 \times 10^{-2}$ m $= 2 \cdot 5 \times 10^{2}$ N C^{-1}, the vertical distance is

$$y = \frac{1 \cdot 6 \times 10^{-19}\,\text{C} \times 2 \cdot 5 \times 10^{2}\,\text{N C}^{-1} \times 1 \cdot 0 \times 10^{-4}\,\text{m}^{2}}{2 \times 9 \cdot 1 \times 10^{-31}\,\text{kg} \times 1 \cdot 0 \times 10^{12}\,\text{m}^{2}\,\text{s}^{-2}}$$

or

$$y = 2 \cdot 2 \times 10^{-3}\,\text{N kg}^{-1}\,\text{s}^{2} = 2 \cdot 2\,\text{mm}$$

Another force field which influences the motion of a charged particle is the magnetic field. Such a field exists in the region between the north and south poles of a magnet or in the vicinity of a wire through which electrical charges are moving. The magnetic force on a particle of charge q moving with speed v perpendicular to the magnetic field is

$$F_{\text{M}} = Bqv \tag{1.20}$$

where B is the magnetic flux density. It is similar to the electric field intensity E in the electrostatic force. The magnetic force acts perpendicular to both the direction of the magnetic field and the motion. It causes the charged particle to move in a circular path when it is in the region of the field as shown in Fig. 1.10. The centripetal force on the particle, which is directed towards the centre of the motion, is provided by the magnetic force, so that from equations 1.15 and 1.20, $Mv^{2}/r = Bqv$ and the radius of the circular path is

$$r = Mv/Bq \tag{1.21}$$

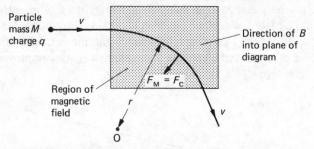

Figure 1.10   The motion of a charged particle in a magnetic field.

1.5   Work, power and energy

An important property of a force is its ability to do work. Work is done when the point of application of a force moves. By definition, work is equal to the product of the force and the distance moved in the direction of the force. If the applied force is F in the x direction and the distance moved is Δx then the work done W is

$$W = F\Delta x \tag{1.22}$$

When F is in newtons and Δx in metres the unit of work is the joule (J). A more general case of an unbalanced force F' applied to a body of mass M at an angle θ to the x direction is shown in Fig. 1.11a. The component of the force acting along the x-axis is $F = F' \cos \theta$ and the corresponding work done when the body moves a distance Δx is

$$W = F\Delta x = F'\Delta x \cos \theta$$

The work done by the component of the force perpendicular to the motion, $F' \sin \theta$, is zero. For this reason the gravitational force $F_G = Mg$ does no work when the object moves in the horizontal plane. The rate at which work is done is described by the power. If a quantity of work ΔW(J) is done in time Δt(s) the power P is

$$P = \Delta W/\Delta t \tag{1.23}$$

where the unit of P is the watt (W = J s^{-1}).

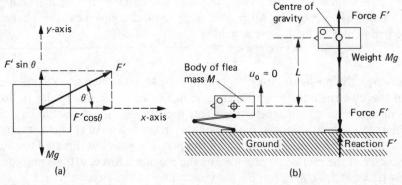

Figure 1.11   (a) The resolution of a force into its components. (b) The initial part of the jump motion of a flea.

Example 1.6

The jump motion of the flea of Fig. 1.2 can be divided into two parts: (a) the initial part where the legs of the flea are exerting a force on the ground, and the body, mass M, is experiencing an upwards acceleration; and (b) the free flight under the influence of the gravitation force which produces an acceleration in the downwards direction. What is the work done and the power developed during the initial part of the jump?

The initial part of the jump is shown in Fig. 1.11b. The legs exert a force F' on the ground which is balanced by an equal and opposite reaction force so that the feet do not move. The centre of gravity of the flea, which is the point at which the force due to gravity may be assumed to act, experiences a net, unbalanced, force $F = F' - Mg$, where Mg is the weight of the flea. The distance L moved by the centre of gravity during this part of the motion is approximately equal to the length of the legs. If the acceleration (in the upwards direction) experienced by the body, mass M, is a then the force on the mass is $F = Ma$ and the work done $W = MaL$. The mass of a typical flea is about 0.20 mg $= 0.20 \times 10^{-6}$ kg. From Fig. 1.2, $a = 1.0 \times 10^3$ m s^{-2} and $L = 0.50 \times 10^{-3}$ m so that

$$W = 0.20 \times 10^{-6} \text{ kg} \times 1.0 \times 10^3 \text{ m s}^{-2} \times 0.50 \times 10^{-3} \text{ m}$$

$$W = 1.0 \times 10^{-7} \text{ J}$$

If the velocity of the flea (in the y direction) during the initial part of the motion is called u, from Equ. 1.4, $(y - y_0) = L = (u + u_0)t/2$. Since $u_0 = 0$ we have $t = 2L/u$ where u is the velocity at the end of this part of the motion. From Equ. 1.23 the power developed $P = W/t = Wu/2L$. Substituting values from Fig. 1.2, $u = 1.0$ m s^{-1}, $L = 0.50 \times 10^{-3}$ m gives

$$P = \frac{1.0 \times 10^{-7} \text{ J} \times 1.0 \text{ m s}^{-1}}{0.50 \times 10^{-3} \text{m}} = 2.0 \times 10^{-4} \text{W}$$

Consider a body of mass M moving in the x direction with speed v_0, (Fig. 1.12a). When it reaches position x_0 a constant force acts on it until it reaches position x. This constant force produces a constant acceleration, $a = F/M$, and the speed of the object at position x is v. From Equ. 1.6, $v^2 = v_0^2 + 2a(x - x_0)$ we have, substituting for a and multiplying both sides of the equation by $M/2$,

$$Mv^2/2 = (Mv_0^2/2) + F(x - x_0) \tag{1.24}$$

Since $(x - x_0)$ is the distance Δx moved under the influence of the force F, the product $F(x - x_0)$ is the work W done by the force. Equ. 1.24 can be re-arranged to give

$$W = (Mv^2/2) - (Mv_0^2/2) = \Delta T \tag{1.25}$$

$Mv_0^2/2$ and $Mv^2/2$ are called the initial and the final kinetic energies of the body. The unit of energy is the joule (J). The symbol commonly used for kinetic energy is T, the work done, W, being equal to the change, ΔT, in the kinetic energy of the body. Kinetic energy is the energy that a body possesses because

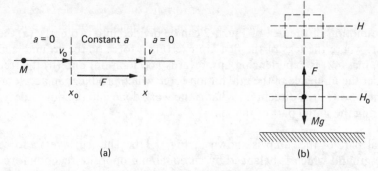

Figure 1.12 The work done by a force in changing: (a) the kinetic energy; and (b) the potential energy of a body.

of its motion and it is a measure of the work required to bring the body to rest.

Another kind of energy is that possessed by a body because of its position. This is called the potential energy U. Fig. 1.12b shows an object of mass M raised from height H_0 to height H by applying a force F equal and opposite to the gravitational force Mg. The work done is $W = F\Delta y = Mg(H - H_0)$ or

$$W = MgH - MgH_0 = \Delta U \qquad (1.26)$$

The quantities MgH_0 and MgH are called the initial and the final potential energies respectively. The work done, W, in raising a body against the gravitational force is equal to the change ΔU in its potential energy.

If a mass M is at rest at height H above some surface it has potential energy MgH and zero kinetic energy (Fig. 1.13a). When released it is accelerated by the gravitational field, in the downwards direction, and gains speed until it hits the surface. At this point its kinetic energy is $Mv^2/2$ and its potential energy is zero. For this motion the changes in the kinetic energy, ΔT, and the potential energy, ΔU, are

$$\Delta T = Mv^2/2 - 0 = Mv^2/2 \qquad (1.27)$$

$$\Delta U = 0 - MgH = -MgH \qquad (1.28)$$

Since the work done by the gravitational force equals the change in the kinetic energy, $W = \Delta T$, we have $F_G H = MgH = Mv^2/2$. From Equ. 1.28, $MgH = -\Delta U$ so the change in the kinetic energy must be equal and opposite to the change in the potential energy:

$$\Delta T = -\Delta U \qquad (1.29)$$

This means that the kinetic energy gained during the fall is equal to the potential energy lost. The total mechanical energy E of an object in the gravitational field is

$$E = T + U$$

E is constant because any change in T is compensated by an equal and opposite change in U. This is the principle of the conservation of mechanical energy which states that the total amount of mechanical energy in a system remains constant although it may change from one form to the other. When the mass is dropped from rest its initial mechanical energy, which is completely potential, is gradually converted to kinetic energy as the speed of the body increases and it loses height. At the end of the fall the mechanical energy is completely kinetic.

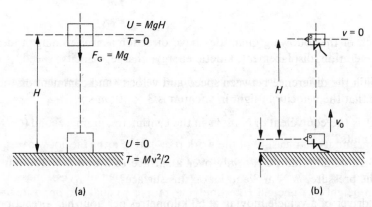

Figure 1.13 (a) The conversion of potential energy into kinetic energy. (b) The free flight of the flea.

Example 1.7

The second part of the jump motion of a flea, the free flight (see Example 1.6), can be analysed using energy methods. If the speed, v_0 ($= u$ of Example 1.6) at the start of this part of the motion is $1 \cdot 0$ m s^{-1} what is the jump height?

From Fig. 1.13b the distance L moved during the first part of the motion is much smaller than H so that it can be ignored and we take H as the height above the surface. Initially $U = 0$ and $T = Mv_0^2/2$. At maximum height H, $U = MgH$ and $T = 0$. Thus $\Delta T = -Mv_0^2/2$, the minus sign indicating that kinetic energy has been lost, and $\Delta U = MgH$. From Equ. 1.29, $MgH = Mv_0^2/2$ so that $v_0^2 = 2gH$ or $H = v_0^2/2g$. Substituting values

$$H = \frac{(1 \cdot 0 \text{ m s}^{-1})^2}{2 \times 9 \cdot 8 \text{ m s}^{-2}} = 51 \text{ mm}$$

as in Example 1.4.

The gravitational force is called a conservative force because the mechanical energy of a body in the gravitational field is constant. Not all forces are conservative. When an object moves across a horizontal surface ($U = $ constant) it gradually slows down indicating a loss in kinetic (and therefore mechanical) energy. This is because of the frictional forces between the object and the surface and the air which impede the motion. Much of the mechanical energy lost is converted into heat energy and the temperatures of the body and surface and the air increase. To maintain constant mechanical energy additional work must be done on the object. This can be provided by pushing the object.

Chemical energy is converted into mechanical energy by the muscles of the arms and the legs which do work on the body and produce mechanical energy of motion and thermal energy.

Conservation of energy is never violated but mechanical energy is not conserved when non-conservative forces are present.

Problems

1.1   Which of the following quantities have direction as well as magnitude: force, acceleration, displacement, kinetic energy, charge, velocity, speed?

1.2   Explain the difference between speed and velocity and comment on the statement that the velocity of light in vacuum is 3×10^8 m s^{-1}.

1.3   What is the equivalent of N m^{-2} s in the centimetre, gram, second (CGS) system of units?

1.4   A force of 20 N acts uniformly over a surface of area $3\cdot0 \times 10^2$ mm^2. What is the pressure, in N m^{-2} and Pa on this surface?

1.5   The driver of a vehicle moving at 60 kilometres per hour has a reaction time of $0\cdot80$ s. If the driver sees a red stop light how far does the vehicle travel before it comes to rest assuming that the brakes produce a constant acceleration of $-3\cdot0$ m s^{-2}? What factors influence the stopping distance?

1.6   The data given below showing the vertical (y) position of a flea at different times were obtained from a high-speed film of the jump motion. (The data shown in Fig. 1.2 were adapted from this so that the equations for constant acceleration could be applied.) Using these data plot a graph of position as a function of time and by taking gradients determine and plot the velocity as a function of time. Is the acceleration constant during the initial part of the jump motion?

Time t/ms	0	0·25	0·50	0·75	1·00	1·25	1·50	1·75	2·0
Position y/mm	0	0·03	0·10	0·30	0·60	0·95	1·25	1·60	1·90

1.7   A climber of mass 70 kg is attached to a rock face by a rope 20 m long. The climber falls from rest and is eventually stopped by the rope. (a) What is the speed of the climber just before the rope takes the impact? (b) If the motion of the climber stops in $0\cdot20$ s what is the acceleration experienced by the climber and the force experienced by the rope? Assume that the acceleration during impact is constant.

1.8   A shot-putter of mass 80 kg moves across the throwing circle at $2\cdot0$ m s^{-1} and throws a $8\cdot0$ kg shot horizontally, transferring all his momentum to the shot and bringing himself to rest. Calculate (a) the speed of the shot on leaving the hand and (b) the average force exerted on the shot by the athlete if his arm straightens in $0\cdot10$ s.

1.9 Most of the cell wall in a suspension of broken bacteria can be separated from the liquid by centrifugation with a centripetal acceleration of $5 \cdot 0 \times 10^3$ g. ($g = 9 \cdot 81$ m s^{-2} is another unit used to describe acceleration.) If the sample cell is 100 mm from the centre of rotation what rotational speed should be used? Express your answer in rad s^{-1} and revolutions per minute.

1.10 A car is driven at high speed around a bend. Explain why bald tyres are dangerous in this situation. On a suitable diagram, indicate all forces acting on the driver and explain his motion relative to the car seat.

1.11 An object weighing 49 N is attached to the free end of a muscle which is hung vertically from a rigid support. When the muscle is stimulated it exerts an unbalanced force of $2 \cdot 5$ N in the upwards direction on the object for a period of 50 ms. Calculate, using equations 1.3 and 1.5, the position, y, and the speed, v, of the mass for different times between zero and 50 ms and plot y–t and v–t graphs.

1.12 The lower jaw of a mammal is acted on by two muscles which produce forces of (a) 230 N in a direction 30 degrees to the $+x$-axis and (b) 260 N in a direction 140 degrees to the $+x$-axis. Draw a diagram showing these forces, resolve them into their x and y components and by adding the components (remembering to take account of the direction) determine the magnitude and direction of the resultant force on the jaw.

1.13 In a rowing racing shell each oarsman exerts a force of $5 \cdot 4 \times 10^2$ N on the handle of the oar which moves a distance $0 \cdot 60$ m during each stroke. If the stroke rate is 34 per minute what is the total power (in watts and horsepower) developed by the eight oarsmen?

1.14 A man pushes a lawnmower over horizontal ground. His arms exert a force of 10 N at an angle of 30 degrees to the horizontal, and the lawnmower moves at a constant speed of $0 \cdot 50$ m s^{-1}. Since a force is exerted by the man, why is the speed constant? Find the work done by the man each minute, and the corresponding horsepower.

1.15 A high-jumper leaps over a bar $2 \cdot 0$ m from the ground. Estimating the appropriate quantities: (a) How far does the centre of gravity of the jumper move in the vertical direction? (b) What is the speed of the jumper as he leaves the ground?

2

Vibrations and Waves

In the previous chapter we considered two particular types of motion, straight line and circular, which are caused by the action of a constant applied force. Another important type of motion is oscillatory, or vibrational, motion in which the particle moves backwards and forwards about some fixed position. This type of motion is called periodic because it repeats itself after a definite period of time. Many natural phenomena are periodic, e.g. the swaying of a tree in the wind, the vibration of the molecules in a solid body, the beating of a heart. The analysis of the vibrational motion gives information about the internal structure and the properties of the system. When a vibrating particle passes its energy to a neighbouring particle a wave is set up which transmits energy from one location to another. Sound waves, which cause the sensation of hearing, consist of vibrating air molecules. Light waves, which cause the sensation of vision, consist of an oscillatory electromagnetic field.

2.1 Simple harmonic motion

The most important type of vibration is simple harmonic motion (SHM). A mathematical description of this motion is easily derived because of its similarity to uniform circular motion.

Consider a point P moving with constant speed v, and angular velocity ω, around the circumference of a circle radius r as shown in Fig. 2.1a. The projection of its position onto the y-axis is the point Q which is distance y from the origin O, $y = OQ$. As the point P moves around the circle the point Q moves backwards and forwards about O. From the triangle OPQ, $y/r = \cos\theta$ so that the position coordinate of Q is $y = r\cos\theta$. Assuming that the line segment OP passes through the y-axis at time $t = 0$ the value of θ at any time t is θ = angular velocity × time = ωt so

$$y = r\cos(\omega t) \tag{2.1}$$

This equation describes the position of the point Q relative to the centre of motion. It is shown graphically in Fig. 2.1b. The function $\cos(\omega t)$ is a periodic function which varies between ± 1. It can have the same value for different values of t. For example $\cos(\omega t) = +1$ when $t = 0$, $2\pi/\omega$, $4\pi/\omega$ etc. The time $T = 2\pi/\omega$ required to complete one cycle is called the period (§ 1.3).

If a particle is at point Q and its motion along the y-axis is given by an equation of the same form as Equ. 2.1 its motion is said to be simple harmonic. The amplitude y_0 of the vibration is the maximum value of the posi-

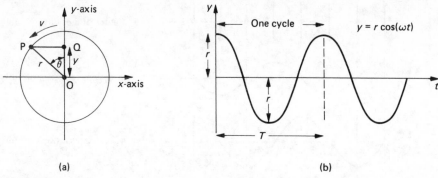

Figure 2.1 (a) The projection of the motion of a point P onto the y-axis. (b) The variation of the position y with time for a particle executing SHM

tion coordinate y (either positive or negative); y_0 is equivalent to r in Equ. 2.1. The frequency f(Hz) of the vibration is $f = 1/T = \omega/2\pi$ so that the position of the particle executing SHM is

$$y = y_0 \cos(2\pi f t) = y_0 \cos(\omega t) \tag{2.2}$$

The velocity $v_y = \Delta y/\Delta t$ of the particle Q can be obtained by taking the gradient of the position–time curve as explained in Section 1.1. Considering Fig. 2.2, at time $t = 0$ the gradient of the y–t curve is zero so $v_y = 0$. As the time increases the gradient becomes negative so that v_y is negative. This means that Q is moving in the $-y$ direction (as in Fig. 2.1b). The gradient of the y–t curve, and hence v_y, is a negative maximum when y is zero. At longer times the velocity becomes less negative, passes through zero when y is a negative maximum, and is then positive. This means that the particle is moving in the $+y$ direction. The variation of the particle velocity as a function of time, which is shown graphically in Fig. 2.2c can be represented by the equation

$$v_y = -y_0 \, \omega \sin(\omega t) \tag{2.3}$$

The maximum values that a sine or cosine function can take are ± 1 so the maximum velocity is $\pm \omega y_0$. This occurs when y is zero, i.e. at the centre of the motion. The velocity is zero when $\omega t = 0, \pi, 2\pi \ldots$ which occurs when y is a positive or negative maximum. Squaring both sides of Equ. 2.3 gives $v_y^2 = y_0^2 \omega^2 \sin^2(\omega t)$. Since $\sin^2(\omega t) + \cos^2(\omega t) = 1$ we have $v_y^2 = y_0^2 \omega^2 [1 - \cos^2(\omega t)]$. From Equ. 2.2, $\cos^2(\omega t) = y^2/y_0^2$ so that

$$v_y^2 = \omega^2(y_0^2 - y^2) \tag{2.4}$$

The acceleration of the particle is the rate of change of velocity with time, $a_y = \Delta v_y/\Delta t$. It is obtained by taking the gradient of the velocity–time curve. The variation of the acceleration during a vibration cycle is shown in Fig. 2.2d. It can be represented by the equation

$$a_y = -y_0 \omega^2 \cos(\omega t) = -\omega^2 y \tag{2.5}$$

The acceleration of a particle executing simple harmonic motion is proportional to its distance y from the centre of motion and is directed towards this point, as indicated by the negative sign. The constant of proportionality

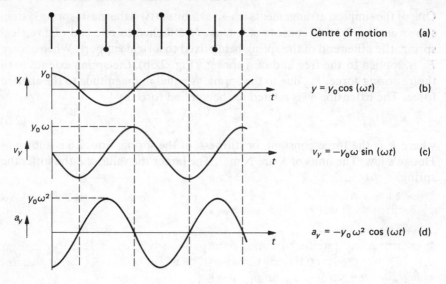

Figure 2.2 (a) The particle executing SHM. (b) The variation of y with t. (c) The variation of v_y with t. (d) The variation of a_y with t.

is equal to the square of the angular frequency ($\omega = 2\pi f$) of the motion. The acceleration has a maximum value when the position coordinate y is a maximum and is zero when $y = 0$.

Example 2.1

The position of a particle executing SHM is given by $y = 3 \cdot 0 \cos(0 \cdot 25\pi t)$ metres. What is the amplitude and frequency of the motion? Find (a) the position, (b) the velocity and (c) the acceleration of the particle after $1 \cdot 0$ s.

Comparing the above equation with $y = y_0 \cos(\omega t)$, the amplitude is $y_0 = 3 \cdot 0$ m. Since $\omega = 2\pi f = 0 \cdot 25\pi$, the frequency $f = 0 \cdot 25\pi / 2\pi = 0 \cdot 13$ s^{-1}.

(a)   When $t = 1 \cdot 0$ s, $y = 3 \cdot 0 \cos(0 \cdot 25\pi)$ metres. $\cos(0 \cdot 25\pi) = \cos(45°) = 0 \cdot 707$ and

$$y = 3 \cdot 0 \text{ m} \times 0 \cdot 707 = 2 \cdot 1 \text{ m}$$

(b)   The velocity is given by Equ. 2.3, with $\sin(0 \cdot 25\pi) = \sin(45°) = 0 \cdot 707$

$$v_y = -3 \cdot 0 \text{ m} \times 0 \cdot 25\pi \text{ s}^{-1} \times 0 \cdot 707$$

$$v_y = -1 \cdot 7 \text{ m s}^{-1}$$

The speed is $1 \cdot 7$ m s^{-1} in the $-y$ direction. Note that the speed can be obtained from Equ. 2.4, but the direction of motion is not specified.

(c)   Putting $\omega = 0 \cdot 25\pi$ s^{-1} and $y = 2 \cdot 1$ m into Equ. 2.5, the acceleration is

$$a_y = -(0 \cdot 25\pi \text{ s}^{-1})^2 \times 2 \cdot 1 \text{ m} = -1 \cdot 3 \text{ m s}^{-2}$$

The negative sign shows the acceleration to be in the $-y$ direction.

One of the simplest arrangements which exhibits SHM is the mass–spring system shown in Fig. 2.3c. It consists of a mass M attached to one end of a vertical spring, the other end of the spring being fixed to a rigid support. When a force F_A is applied to the free end of a spring, (Fig. 2.3b), the spring extends until the restoring force, F_R, due to the spring, is equal in magnitude to the applied force. The extension y' is related to the applied force by

$$F_A = Ky' \tag{2.6}$$

where K is the force constant, or stiffness, of the spring. Equ. 2.6 is known as Hooke's law. The units of K are N m^{-1}. The larger the value of K the stiffer the spring.

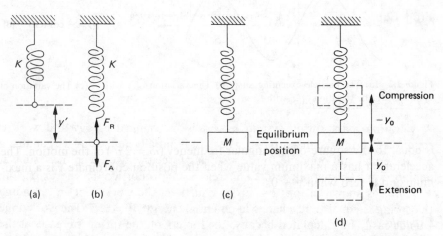

Figure 2.3 (a) The unstretched spring. (b) Extension of the spring due to an applied force F_A. (c) The mass – spring system at equilibrium. (d) Simple harmonic motion of the mass about the equilibrium position.

When a mass M is attached to an unstretched spring, $F_A = Mg$, and the spring extends until $|F_R| = Mg$. The mass then remains stationary. This is called the equilibrium position. If the mass is now displaced, in the vertical direction, the motion can be analysed by considering the unbalanced forces and the motion described relative to the equilibrium position.

If the mass is pulled downwards, in the positive y direction, by an amount y, measured relative to the equilibrium position, and then released it experiences an unbalanced force $F = -Ky$ due to the restoring force of the spring. The minus sign indicates that this force is in the upwards direction. From Newton's second law, $F = -Ky = Ma_y$, so that the acceleration of the mass is

$$a_y = -(K/M)y$$

Comparison with Equ. 2.5 shows that the mass executes SHM about the equilibrium position with $\omega = \sqrt{K/M}$. The amplitude, y_0, of the motion, which is shown in Fig. 2.3d, is the initial displacement from the equilibrium position. The position y of the mass is given by Equ. 2.2 which is shown graphically in Fig. 2.2b. The frequency of the vibration, which is called the natural frequency,

f_n, of the mass–spring system, is

$$f_n = \frac{1}{2\pi}\sqrt{\frac{K}{M}} \tag{2.7}$$

The natural frequency increases with increasing stiffness and decreasing mass.

Example 2.2

The head–neck–shoulder system of a person may be considered as a mass M (the head), a spring K (the neck) and a rigid foundation (the shoulders). The natural frequency of the head–neck system is 30 Hz. If the mass of the head is 5·5 kg what is the stiffness K of the neck?

From Equ. 2.7, $K = 4\pi^2 f_n^2 M$, substituting values gives

$K = 4 \times 9\cdot9 \times 9\cdot0 \times 10^2\,\text{s}^{-2} \times 5\cdot5\,\text{kg}$

$K = 2\cdot0 \times 10^5\,\text{N m}^{-1}$

2.2 Energy of vibration

A vibrating mass–spring system possesses mechanical energy and work is required to bring the mass to rest. The mass is initially pulled down by applying a force equal and opposite to the restoring force trying to keep the system at equilibrium. This applied force increases in proportion to the extension, $F = Ky$, as shown in Fig. 2.4. From Equ. 1.26, the work done in extending the spring is equal to its change in potential energy, $W = \Delta U$. The work done cannot easily be calculated because the force F on the spring increases as the point of application, y, changes. However the extension of the spring can be divided into a number of small parts Δy within which the force may be assumed to be constant. During the first extension increment, from Equ. 1.22, the work

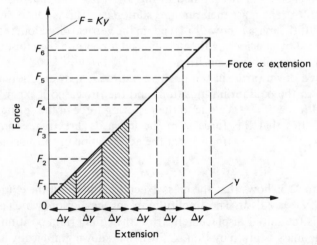

Figure 2.4 The calculation of the potential energy stored in an extended spring.

done is $\Delta W_1 = F_1 \Delta y$ which is equal to the first shaded area in Fig. 2.4. The work done during the second extension increment is $\Delta W_2 = F_2 \Delta y$ which is equal to the second shaded area. The total work done in extending the spring by an amount y is $W = \Delta W_1 + \Delta W_2 + \Delta W_3 + \ldots = F_1 \Delta y + F_2 \Delta y + F_3 \Delta y + \ldots$ which is equal to the area between the line $F = Ky$ and the y-axis from zero to y. This is $(\frac{1}{2})(Ky \times y) = Ky^2/2$. Defining the initial potential energy as zero, the potential energy of the system when the mass is at position y is

$$U = Ky^2/2 \qquad (2.8)$$

It has a maximum value $Ky_0^2/2$ when $y = y_0$ and is zero at the equilibrium position when $y = 0$.

When the mass M is moving with speed v_y the kinetic energy of the system is $T = Mv_y^2/2$. Substituting for v_y from Equ. 2.4 gives

$$T = M\omega^2(y_0^2 - y^2)/2 \qquad (2.9)$$

so that the kinetic energy has a maximum value $M\omega^2 y_0^2/2$ when $y = 0$ and is zero when $y = y_0$. The total mechanical energy E of the vibrating system is $E = T + U$. From equations 2.8 and 2.9, and substituting $K = M\omega^2$ (from Equ. 2.7) gives

$$E = (K/2)(y_0^2 - y^2) + (K/2)y^2 = (K/2)y_0^2 \qquad (2.10)$$

Thus the total mechanical energy of a vibrating system is constant and depends only on the stiffness of the spring and the vibration amplitude. However there is an interchange between the kinetic and potential energies during the vibration cycle. This is shown in Fig. 2.5. When the mass is at rest, at the lower limit of the vibration, the mechanical energy is completely potential and is stored in the spring. When the mass is released its speed increases under the acceleration of the restoring force of the spring and it gains kinetic energy. The gain in the kinetic energy is equal to the potential energy lost by the spring. At the centre of motion all of the potential energy has been converted to kinetic energy. At this position the mass is moving at maximum speed and it overshoots the equilibrium position. The spring is compressed and exerts a retarding force on the mass so that the mass loses kinetic energy and the spring gains potential energy. Eventually the mass stops moving, at the upper limit of the vibration, and once again the mechanical energy is completely potential.

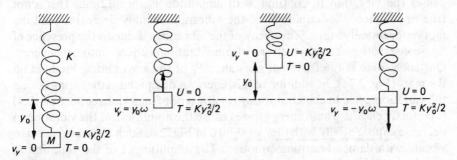

Figure 2.5 The interchange of kinetic and potential energies during the vibration of a mass – spring system.

The restoring force due to the spring then accelerates the mass downwards so that potential energy is transformed into kinetic energy and the process repeats itself.

Example 2.3

A spring extends by $5 \cdot 0 \times 10^{-2}$ m when a mass of $0 \cdot 10$ kg is hung from it. What is the force constant of the spring? The mass is pulled down a further $0 \cdot 10$ m and released. Calculate the vibrational energy of the system and the distance of the mass from its equilibrium position when its kinetic and potential energies are equal.

When the mass is hung from the spring, the gravitational force $F_G = Mg$ extends the spring a distance y' downwards until the restoring force in the spring $F_R = Ky'$ (acting upwards) balances F_G. This is the equilibrium condition as no unbalanced force acts on the system. Since at equilibrium $Ky' = Mg$, $K = Mg/y'$ and substituting $y' = 5 \cdot 0 \times 10^{-2}$ m when $M = 0 \cdot 10$ kg we have

$$K = \frac{0 \cdot 10 \text{ kg} \times 9 \cdot 8 \text{ m s}^{-2}}{5 \cdot 0 \times 10^{-2} \text{ m}}$$

or

$$K = 20 \text{ N m}^{-1}$$

Using Equ. 2.10, where $y_0 = 0 \cdot 10$ m, the vibrational energy is

$$E = (20 \text{ N m}^{-1}/2) \times (0 \cdot 10 \text{ m})^2 = 10 \times 10^{-2} \text{ J}$$

The kinetic and potential energies of the mass are equal when $(K/2)(y_0^2 - y^2) = (K/2)y^2$. This occurs when $y = y_0/\sqrt{2}$ and since $y_0 = 0 \cdot 10$ m

$$y = 0 \cdot 10 \text{ m}/\sqrt{2} = 7 \cdot 1 \times 10^{-2} \text{ m}$$

2.3 Damping and resonance

In the previous section we assumed that there are no energy losses in the vibrating system so that the interchange between kinetic and potential energy causes the vibration to continue with amplitude y_0 for all time. This is not true in practice, the amplitude of the vibration steadily decreases until the motion eventually stops. The decay of the vibrations is due to the presence of non-conservative forces which transform vibrational energy into heat energy. One such force is the frictional resistance F_F of the surrounding medium on the mass (Fig. 2.7a). In addition to this, energy is dissipated in the spring during the vibration cycle because no spring (or any type of material, see Section 10.4) is perfectly elastic. These energy losses cause the amplitude of the vibration to decrease exponentially with time as shown in Fig. 2.6. Such a system is said to vibrate with damped harmonic motion. The amplitude A of the vibration at any time t is given by $A = A_0 e^{-(f\Phi)t}$ where A_0 is the amplitude at zero time and

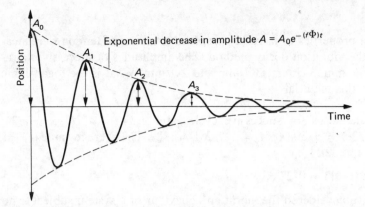

Figure 2.6 The exponential decay of vibrations, with frequency f, in damped harmonic motion.

Φ is called the logarithmic decrement. This is defined as the natural logarithm of the ratio of the amplitudes of successive vibration cycles:

$$\Phi = \log_e(A_0/A_1) = \log_e(A_1/A_2) = \log_e(A_2/A_3) \tag{2.11}$$

The degree of damping, and hence the logarithmic decrement, depends upon the medium in which the system vibrates and the amount of energy dissipated by the spring. For example, the damping of a mass–spring system vibrating in water is much larger than when it vibrates in air (Fig. 2.7a). The vibration behaviour of systems with different degrees of damping is shown in Fig. 2.7b. In curve (i) the system is underdamped and the mass vibrates about its equilibrium position. If the damping is increased to some critical value, curve (ii), the mass moves more slowly towards its equilibrium position but does not oscillate. With overdamped systems, curve (iii), the mass takes a long time to reach equilibrium. Critical damping is employed in many measuring instruments which use an indicating needle and calibrated scale, e.g. voltmeters, ammeters, speedometers, weighing scales (see § 6.5). An underdamped pointer would oscillate about its final position whereas an overdamped system would move so slowly that the quantity to be measured may have changed before the reading can be taken.

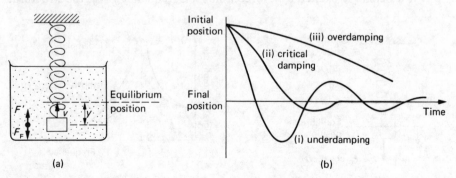

Figure 2.7 (a) A highly damped vibrating system. (b) The vibration behaviour for different degrees of damping.

Example 2.4

The damping properties of a freshly prepared specimen of tendon were measured using the vibration decay method. The amplitudes of successive vibrations were 4·7 mm, 3·5 mm, 2·7 mm and 2·1 mm. What is the logarithmic decrement of this material?

The amplitude ratios for successive cycles are: $A_0/A_1 = 4·7/3·5 = 1·34$, $A_1/A_2 = 3·5/2·7 = 1·30$, $A_2/A_3 = 2·7/2·1 = 1·29$. The average value is 1·31 so that from Equ. 2.11

$$\Phi = \log_e(1·31) = 0·27$$

So far we have considered the vibration behaviour of a system subject to no external forces. This is called free vibration. The frequency f_n of the free vibration is governed by the properties, i.e. the stiffness and the mass, of the system. A system can also be set into vibration by an externally applied force which varies periodically in time, e.g. $F = F_0 \cos(2\pi ft)$ where F_0 is the amplitude and f the frequency of the exciting force. This is called forced vibration and is shown in Fig. 2.8a. The externally applied force alternately extends and compresses the spring and the system undergoes forced harmonic motion with frequency f.

For an undamped system the displacement y of the mass is given by $y = [(F_0/K)/\sqrt{1 - (f/f_n)^2}] \cos(2\pi ft)$ where $f_n = (1/2\pi)\sqrt{K/M}$. As the excitation frequency f approaches the natural frequency, f_n, of the system, the vibration amplitude $y_0 = (F_0/K)/\sqrt{1 - (f/f_n)^2}$ increases and the energy of vibration increases (see Equ. 2.10). When $f = f_n$ the vibration amplitude passes through a maximum value. This is called resonance. A resonance curve is shown in Fig. 2.8b. For an undamped system the resonance frequency is the same as the natural frequency f_n whereas for a damped system it is less than f_n. The shape of the resonance curve depends upon the degree of damping, the smaller the damping the sharper the resonance.

The human body consists of a complex arrangement of structural elements which have mass, stiffness and damping. Each element has a characteristic natural frequency. Some of the body resonance frequencies are given in Table 2.1. If a person is excited by a force which has frequency $f = f_n$ the structural

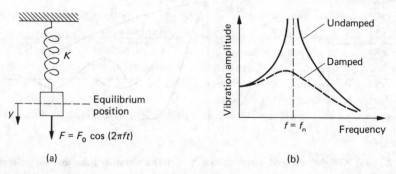

Figure 2.8 (a) The forced vibration of a mass–spring system. (b) Resonance curves for damped and undamped systems.

element goes into resonance vibration. When the amplitude of the vibration is large tissue damage may occur. In the case of the eyeball resonance there is a decrease in visual acuity. Body excitation at these frequencies is common in transport vehicles.

Table 2.1 Body resonance frequencies.

System	f/Hz	System	f/Hz
Abdominal mass	4–8	Head/neck	30
Spine/trunk	7	Eyeball	80

2.4 Wave motion

A wave is a periodic disturbance which moves from one location to another. The way in which a mechanical wave is generated and propagated along a wire is shown in Fig. 2.9. One end of the wire is attached to the tip of a bar which can be displaced in the y direction. When the bar is displaced upwards by an amount y_0 and returned to its equilibrium position the small section of wire attached to the tip moves up and down in the same way. This vertical disturbance is transmitted to adjacent sections of wire and a displacement pulse is transmitted along the wire in the z direction, (Fig. 2.9a). The speed of propagation of the pulse is v so that it takes a time $t = z/v$ for it to travel a distance z.

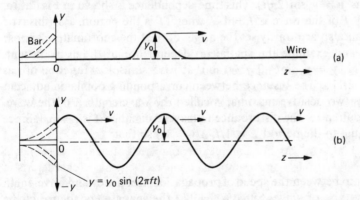

Figure 2.9 (a) The propagation of a pulse in a wire. (b) The generation and propagation of a mechanical wave in a wire.

If the bar vibrates with SHM such that the displacement y of the tip is $y = y_0 \sin(2\pi ft)$ the displacement of the section of wire attached to the bar is given by the same equation. This disturbance is transmitted to adjacent sections of the wire and a mechanical wave of amplitude y_0 is set up. This is shown in Fig. 2.9b. Since it takes a time $t = z/v$ for the disturbance to travel a distance z along the wire the displacement y at any point z lags behind that at the origin, $z = 0$, and

$$y = y_0 \sin[2\pi f(t - z/v)] \tag{2.12}$$

It should be remembered that the small sections of wire do not move in the direction of propagation z. They move up and down in the vertical direction like a mass on a spring executing SHM. This is shown in Fig. 2.10a. The wave is propagated by the transfer of energy of vibration from one section of wire to another. In a water wave travelling in the z direction the particles execute uniform circular motion as shown in Fig. 2.10b and there is no net horizontal movement (except close to the shore).

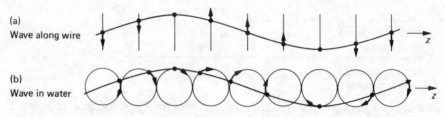

(a)
Wave along wire

(b)
Wave in water

Figure 2.10   (a) The motion of segments of a wire. (b) Particle motion in a water wave.

The size of the disturbance in a wave depends upon time and the distance from the origin. The time and space dependence of the disturbance can be examined separately. Consider a steady wave in which the amplitude is constant, independent of t and z. Since the wave amplitude is constant and the disturbance is periodic, both in t and z, we can define any point on the axis as being the origin, $z = 0$. Substituting $z = 0$ into Equ. 2.12 gives the displacement of the string as $y = y_0 \sin(2\pi ft)$. This time dependence is shown in Fig. 2.11a. The frequency f of the wave is $f = 1/T$ where T is the period. For this type of wave we can also arbitrarily set the origin of the time coordinate. Suppose we set $t = 0$ and examine the spatial (z) dependence of the displacement. From Equ. 2.12, $y = y_0 \sin(-2\pi fz/v)$ so that y is a sinusoidal function of z as shown in Fig. 2.11b. The distance between corresponding points on adjacent cycles, such as two positive maxima, is called the wavelength λ of the wave. The time t required for the disturbance to travel a distance λ is $t = \lambda/v$. Since this time is equal to the period $T = 1/f$, $\lambda/v = 1/f$ so that

$$v = f\lambda \tag{2.13}$$

This relationship between the speed of propagation, frequency and wavelength applies to any type of wave. Since $f/v = 1/\lambda$ the spatial dependence of the disturbance in a wave can be written as

$$y = y_0 \sin(kz) \tag{2.14}$$

where the propagation constant $k = 2\pi/\lambda$.

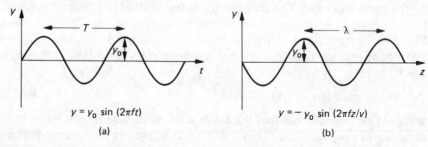

$y = y_0 \sin(2\pi ft)$

(a)

$y = -y_0 \sin(2\pi fz/v)$

(b)

Figure 2.11   (a) The time and (b) the spatial dependence of the disturbance in a wave.

When comparing waves propagating through the same medium the important quantities are the frequency, the amplitude, the wavelength and the time
or space relationship between the waves. Consider two waves which have the
same f, λ and y_0 as shown in Fig. 2.12. If at some position z the disturbance
in each wave does not have the same value at some instant in time we say
that there is a phase difference between the waves. In Fig. 2.12a the disturbance
in wave 1 at $t = 0$ is the same as in wave 2 at time $t = T/2$. Since one period
is equivalent to $2\pi fT$ radians $= 2\pi$ radians ($= 360$ degrees) there is a phase
difference of π radians (or 180 degrees) between wave 1 and wave 2. If the
equation for wave 1 is $y_1 = y_0 \sin(2\pi ft)$, then wave 2 can be represented by
$y_2 = y_0 \sin(2\pi ft + \phi)$ where ϕ is the phase angle, which in this case is π radians.
Since $\sin(2\pi ft + \pi) = -\sin(2\pi ft)$ the displacement is $y_2 = -y_0 \sin(2\pi ft)$ as
shown in the diagram.

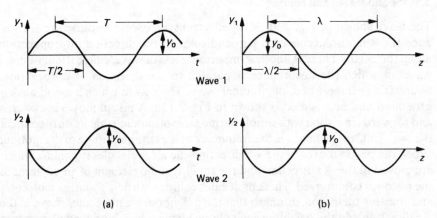

Figure 2.12 Two waves, y_1 and y_2, with (a) a phase difference and (b) a spatial difference.

Two waves with a spatial difference $\delta = \lambda/2$ are shown in Fig. 2.12b. This
situation is identical to that of Fig. 2.12a so that a spatial shift of $\lambda/2$ is the
same as a phase shift of π radians. Thus spatial differences and phase differences are equivalent and since a phase difference $\phi = 2\pi$ radians is equivalent
to a spatial difference $\delta = \lambda$, $\phi/2\pi = \delta/\lambda$ or

$$\phi = 2\pi\delta/\lambda \tag{2.15}$$

Example 2.5

The spatial variation of a wave is $y_1 = y_0 \sin(4\pi z)$ metres. What is its wavelength? A second wave of the same amplitude, frequency and wavelength lags
behind the first wave by one-sixth of a wavelength. Find the phase difference
between the waves and write down the equation of the second wave.

Comparing with Equ. 2.14, the propagation constant $k = 4\pi = 2\pi/\lambda$ or $\lambda =
2\pi/4\pi = 0\cdot5$ m.
Using Equ. 2.15 with $\delta = \lambda/6$, the phase difference between the waves is

$$\phi = \frac{2\pi\lambda \text{ m}}{6\lambda \text{ m}} = \pi/3 \text{ radians } (= 60 \text{ degrees})$$

The second wave is described by $y_2 = y_0 \sin(4\pi z - \phi)$ metres, the negative sign indicating that it lags behind the first wave. With $\phi = \pi/3$ radians, we have

$$y_2 = y_0 \sin(4\pi z - \pi/3) \text{ metres}$$

A wave is propagated by the transfer of vibrational energy between the particles of the medium through which it travels. The total energy in a wave is described by its intensity I (W m^{-2}) which is the amount of energy passing perpendicularly through unit area per second. The intensity is directly proportional to the wave amplitude squared, $I \propto y_0^2$, the constant of proportionality depending on the type of wave. This square law means, for example, that a water wave 3 m high has nine times the intensity of a wave 1 m high.

2.5 Sound waves and hearing

The wave shown in Fig. 2.9b is called a transverse wave because the disturbance (i.e. the displacement) is perpendicular to the direction of propagation along the z-axis. There is another important class of wave, called a longitudinal wave, in which the disturbance is parallel to the direction of propagation. Sound is an example of a longitudinal wave. The way in which a sound wave is generated and propagated is shown in Fig. 2.13a. A piston moves backwards and forwards in a tube with simple harmonic motion such that its displacement is $z = z_0 \sin(2\pi ft)$ where f is the frequency and z_0 the amplitude of the motion. When the piston moves in the $+z$ direction the air molecules is contact with it are moved in the $+z$ direction and the small volume element of air adjacent to the piston is compressed. These molecules collide with neighbouring molecules and transfer their momentum so that the neighbouring molecules move in the $+z$ direction and the neighbouring volume element is compressed. A pressure pulse is transmitted down the tube by the transfer of momentum between colliding molecules. When the piston moves in the $-z$ direction the volume element adjacent to the piston is expanded so that pressure is below ambient. This expansion is transmitted down the tube in the same way as the compression. A sound wave is a pulsating compression and expansion which is pro-

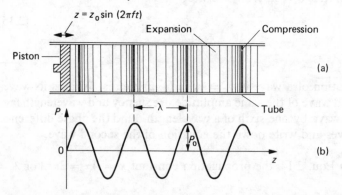

Figure 2.13 (a) The generation and propagation of a sound wave in a tube. (b) Variation of the pressure in a sound wave with position.

pagated through a medium by means of molecular collisions. The pressure P in a sound wave is a sinusoidal function of the distance z from the origin, as shown in Fig. 2.13b and can be represented by the equation $P = P_0 \sin[2\pi f(t - z/v)]$ where f is the frequency (or pitch) of the sound. The speed of propagation v depends upon the medium through which the wave is travelling. For air at normal temperature and pressure ($T = 273 \cdot 15$ K, $P = 1 \cdot 013 \times 10^5$ N m^{-2}), $v = 344$ m s^{-1}. In water the speed of propagation of sound is about $1 \cdot 4 \times 10^3$ m s^{-1}. A pure tone is made up of waves which have the same frequency. Most natural sounds contain waves of different frequencies and amplitudes. The speed (v) is related to the frequency (f) and the wavelength (λ) by Equ. 2.13.

The ear is the sensory organ that detects sound waves. Its function is to transform the pressure fluctuations in a sound wave into electrical impulses. It consists of three parts, the outer, middle and inner ear as shown in Fig. 2.14. The incident sound waves are directed onto the ear drum by the ear canal. The drum is a stiff membrane which is flexible around its periphery. The pressure fluctuations in the sound wave cause the membrane to vibrate. This mechanical vibration is transmitted to the inner ear by means of three bones which are known collectively as the ossicles. The inner ear is made up of several liquid-filled canals inside the temporal bone of the skull. The semicircular canals are the position sensors used in the balance mechanism and the cochlea contains the auditory sensors. The cochlea is a coiled tubular canal filled with a liquid called lymph. The canal is divided into two parts by the basilar membrane which extends from the base of the cochlea almost to the apex of the spiral. The auditory sensor is a system of thin hairs, called the auditory hairs, which are contained within the organ of Corti. This rests on the basilar membrane. When deformed, the auditory hairs produce a series of electrical pulses which are transmitted to the brain by the auditory nerve and cause the sensation of hearing. The ossicles, which are situated in the air-filled middle ear cavity, are connected to the cochlea by the oval window. When the mechanical vibration of the ossicles is transmitted into the liquid filling the cochlea, the basilar membrane vibrates so that the auditory

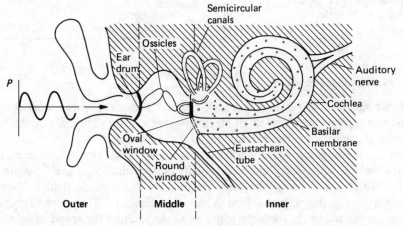

Figure 2.14 A diagram of the peripheral hearing mechanism showing the outer and middle ear which are air filled and the inner ear which is liquid filled.

hairs are deformed and electrical pulses are produced. The stiffness K and the mass M of the basilar membrane varies along its length so different regions of the membrane resonate at different frequencies ($f_n = (1/2\pi)\sqrt{K/M}$). The point of origin of the pulses in the organ of Corti gives an approximate indication of the frequency (or pitch) of the incident sound wave. The final frequency analysis of the sound is done by the brain.

The ear responds to fluctuations in air pressure. A sound wave, in which the pressure fluctuations are of the form $P = P_0 \sin(2\pi ft)$ is usually described in terms of the root-mean-square (RMS) pressure level $p = P_0/\sqrt{2}$ where P_0 is the amplitude of the wave. The smallest RMS pressure p_0 at 1 kHz that can be detected by normal young people is about 2×10^{-5} N m^{-2}, which can be compared with atmospheric pressure, which is about 1×10^5 N m^{-2}. The intensity of a sound (i.e. the energy in the waves) is proportional to p^2. It is found that the ear is sensitive to relative changes in the intensity rather than the absolute value so that it is convenient to use a relative (or logarithmic) scale to describe the intensity of a sound. The sound pressure level (SPL), which has units of decibel (dB), is defined as

$$\text{SPL} = 20 \log_{10}(p/p_0)$$

where p is the RMS pressure in the sound wave and $p_0 = 2 \times 10^{-5}$ N m^{-2} is the threshold of hearing at 1 kHz. The normal range of hearing is from about 20 Hz to 20 kHz, the upper frequency limit decreasing with age or continual exposure to loud noises (greater than 90 dB). The ear is most sensitive for sound of frequency about 3·5 kHz, which is at the centre of the frequency spectrum of speech sounds. The sensitivity is much less at high and low frequencies. Some typical values of SPL are given in Table 2.2.

Table 2.2 Typical sound pressure levels (SPL).

Environment	SPL/dB	Environment	SPL/dB
Threshold of hearing	0	Speech	60–70
Bedroom at night	20–30	Workshop	80–100
Library	30–40	Threshold of pain	140

2.6 Electromagnetic waves

Light waves, radio waves, heat waves, X-rays and gamma rays are examples of a particular type of wave called an electromagnetic wave. The only difference between these waves is their frequency and their origin.

The electromagnetic wave is a transverse wave. It has two components: an electric field E which oscillates sinusoidally in time and space and a magnetic field B which oscillates at the same frequency. The electric field is always perpendicular to the magnetic field as shown in Fig. 2.15. The speed of propagation c is related to the frequency by $c = f\lambda$. In vacuum the speed of light (or any electromagnetic radiation) is $c = 3·0 \times 10^8$ m s^{-1}.

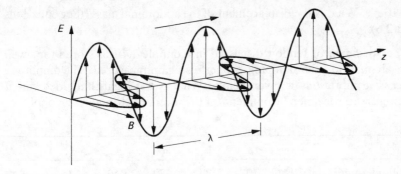

Figure 2.15 The electromagnetic wave.

In the mechanical waves discussed previously the wave propagates by the interchange of energy between vibrating particles. With electromagnetic waves the equivalent interchange is between the electric and magnetic fields. A change in E produces a change in B which itself produces a change in E. Unlike the transverse wave in a wire no material particles move in the path of an electromagnetic wave and for this reason they are one of the waves that are able to travel through a vacuum. The eye is sensitive to light because the retina responds to the electric field. This is discussed in Section 4.3. The way in which the image is formed on the retina is described in Section 7.4.

Problems

2.1 The position coordinate of a particle executing SHM is $y = 5.0 \times 10^{-2} \sin(2\pi t)$ metres. What is the frequency f and the period T of the vibration? Using tables calculate and plot on graph paper the variation of y as a function of t, in 0.1 s intervals for one complete cycle. (Note: (a) the quantity $2\pi t$ is in radians, it must be converted to degrees. (b) $\sin(\theta + \pi/2) = \cos\theta$; $\sin(\theta + \pi) = -\sin\theta$; $\sin(\theta + 3\pi/2) = -\cos\theta$).

2.2 A body of mass 0.10 kg executes SHM of amplitude 0.10 m and period 0.20 s. What is the maximum force acting on it?

2.3 Draw a mechanical analogue consisting of a mass, spring and foundation for the head–neck–shoulder system described in Example 2.2

2.4 When a person is subjected to whole-body vibration his head vibrates such that its position coordinate $y = 5.0 \times 10^{-3} \sin(190\, t)$ metres. (a) Draw a diagram showing how y varies with t. (b) What is the amplitude of the vibration? (c) What is the period of oscillation and frequency f? (d) What is the maximum acceleration experienced by the head? Express your answer in m s^{-2} and g's.

2.5 The abdominal mass sytem can be represented by a mass $M = 4.0$ kg and a single Hookean spring of stiffness K. If the resonance frequency of the abdomen system is 7.0 Hz what is K? A person is subject to a uniform acceleration of 10 g's. What is the displacement of the abdominal mass? (See equations 1.7 and 2.6).

2.6   The specimen of tendon discussed in Example 2.4 was stored for seven days in alcohol and the damping properties again measured. The amplitude of successive vibrations is given below. What is the logarithmic decrement of the preserved specimen? Comment on the effect of storage in alcohol.

	A_0	A_1	A_2	A_3	A_4
Amplitude/mm	2·10	1·90	1·70	1·50	1·35

2.7   By noting that the time elapsed during the decay of a free vibration is $t = mT$ where $m(= 0, 1, 2, \ldots)$ is the number of cycles and T is the period, show that after m vibration cycles the amplitude is $A = A_0 \, e^{-m\Phi}$. Using the data of Problem 2.6 plot $\log_e A$ as a function of m and hence show that the decay is exponential. What is the gradient of the line?
($\log_e A = \log_e A_0 - m\Phi$).

2.8   Describe clearly the similarities and differences between transverse and longitudinal waves. Give examples of each type.

2.9   Consider two waves that can be represented by the equations $y_1 = y_0 \sin(2\pi ft)$ and $y_2 = y_0 \sin(2\pi ft + \pi)$ as shown in Fig. 2.12a. Show using algebra and by geometry that when these waves are added together the total disturbance $y = y_1 + y_2$ is zero at all times.

2.10   What is (a) the RMS value and (b) the amplitude of the pressure in a sound wave which has an SPL of 90 dB.

2.11   How long does it take (a) sound and (b) light to travel a distance of one mile in air? The speed of light in air is approximately the same as that in vacuum.

2.12   Bats navigate by emitting a series of short pulses of sound waves and detecting the sound reflected by objects in their flight path. The frequency of the sound is about 50 kHz and the pulse duration 2·0 ms. If the speed of sound is $3·4 \times 10^2$ m s^{-1} how long does it take for the pulse to make the round trip from bat to object if the object is (a) 1·0 m and (b) 0·20 m from the bat? Explain why the bat decreases the pulse duration as it approaches an object.

2.13   Two machines in a workshop each produce a sound pressure level (SPL) of 80 dB. What is the total SPL when these machines operate at the same time? (Note: you cannot simply add the SPL of each machine, the total RMS pressure $p^2 = p_1^2 + p_2^2$.)

2.14   The hearing threshold (HT) of different groups of workers exposed to high noise levels for different periods of time was measured using an audiometer. From the data given below plot graphs showing the hearing thresholds HT (dB) as a function of the logarithm of frequency of the sound for the different groups.

Frequency/kHz	0·25	0·5	1·0	2·0	3·0	4·0	6·0
HT/dB no exposure	0	2	0	2	2	4	3
HT/dB 15 years	5	5	8	15	40	45	30
HT/dB 50 years	15	15	25	45	50	50	40

The intelligibility of speech is governed by the ability to hear high frequency sounds and the loudness by the ability to hear low frequency sounds. Comment on the ability of each group to hear and understand conversational speech.

3

Atomic and Molecular Structure

Atoms and their combinations as molecules are the basic blocks from which all materials are built. The discovery of the electron, by separating this very small particle of mass $9{\cdot}11 \times 10^{-31}$ kg and charge $-1{\cdot}60 \times 10^{-19}$ C from the atom, led to speculation as to the internal structure of these small atomic building blocks (which have diameter approximately 10^{-10} m). Atoms have been shown to comprise a heavy, positively charged, nucleus around which the much lighter, negatively charged electrons orbit in much the same way as the planets in the solar system orbit the sun. The positive charge on the nucleus is equal to the total negative charge carried by the Z atomic electrons so that an atom is electrically neutral (i.e. the sum of the positive and negative electrical charges is zero). Z is called the atomic number of the atom. The precise nature of the electronic structure of atoms governs how they interact to form molecules and the determination of this structure plays an important role in the analysis of materials.

3.1 The discreteness of atomic electron states

If the atoms in a gas are excited, by passing an electric current through the gas or by heat, light waves are emitted. The wavelength of the light is related to the frequency by

$$c = f\lambda \tag{3.1}$$

where c is the speed of light. In vacuum $c = 3{\cdot}00 \times 10^8$ m s^{-1}. When the light emitted by excited atoms is analysed it is found that emission does not occur at all wavelengths but only at certain discrete wavelengths. This is called the emission spectrum of the gas. Atomic hydrogen emits at three visible wavelengths (red, green and blue) and the visible spectrum consists of three coloured lines as shown in Fig. 3.1. These lines are part of a family of emission lines, called the Balmer series, the wavelengths of which are given by the empirical equation

$$\frac{1}{\lambda} = R\left(\frac{1}{2^2} - \frac{1}{n^2}\right) \tag{3.2}$$

where R is a constant ($= 1{\cdot}097 \times 10^7$ m^{-1}) and n is an integer equal to 3 for the red line, 4 for the green and 5 for the blue. The series includes other lines corresponding to higher values of n, but the eye is insensitive to these wavelengths.

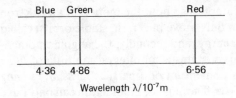

Figure 3.1   The visible emission spectrum of hydrogen.

Any description of the atom must explain this discreteness of the wavelengths of radiation emitted by atoms.

The problem in considering the electrons moving in orbits around the nucleus is that a charged particle in such an orbit should lose energy continuously by the radiation of light and should eventually spiral into the nucleus as shown in Fig. 3.2a. The stability of everyday atoms contradicts this. Such atoms radiate only when energy is given to them and the wavelengths emitted are well defined and characteristic of each atom.

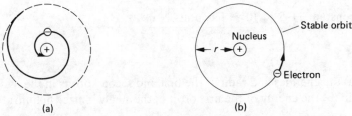

Figure 3.2   (a) Classical and (b) quantum electron orbits.

Niels Bohr attempted to explain this using quantum theory, in which light is considered to be emitted and absorbed in small packets, or quanta, each quantum having a certain energy E which is proportional to the frequency f of the light,

$$E = hf \tag{3.3}$$

where h is Planck's constant equal to $6 \cdot 63 \times 10^{-34}$ J s. Furthermore, each quantum of radiant energy hf does not spread out equally in all directions like the ripples on a pond, but travels in one direction, like a particle. Such a light particle is called a photon.

Example 3.1

The red line in the emission spectrum of hydrogen has a wavelength of $6 \cdot 56 \times 10^{-7}$ m. What is the energy of this radiation?

From equations 3.1 and 3.3, $E = hf = hc/\lambda$ so that

$$E = \frac{6 \cdot 63 \times 10^{-34} \text{ J s} \times 3 \cdot 00 \times 10^{8} \text{ m s}^{-1}}{6 \cdot 56 \times 10^{-7} \text{ m}}$$

$$E = 3 \cdot 04 \times 10^{-19} \text{ J}$$

The Bohr theory makes certain assumptions about the motion of an electron in the atom. Firstly, the electron can only move in certain stable orbits, called stationary states, without radiating energy, and secondly, the angular momentum of the electron in these stationary states is an integral multiple of $h/2\pi$ units. From Equ. 1.13, the angular momentum of a particle mass M, moving with speed v in a circle of radius r is Mvr. The angular momentum can only have certain definite values, or is quantized, such that $Mvr = nh/2\pi$. We call n the principal quantum number, and this is a convenient description of the electron energy levels in the atom. Application of the theory gives the energy of an electron in a state of principle quantum number n as

$$E_n = -RhcZ^2/n^2 \tag{3.4}$$

where Z is the atomic number of the atom and the negative sign indicates that the electron is bound to the atom and energy must be supplied to remove it. The corresponding radius of the nth orbit is

$$r_n = \frac{11 \cdot 5 \times 10^{-29} n^2}{ZRhc} = -\frac{11 \cdot 5 \times 10^{-29} Z}{E_n} \tag{3.5}$$

where the constant $11 \cdot 5 \times 10^{-29}$ has units N m^2.

Example 3.2

Calculate (a) the energies and radii of the first and second orbits in the hydrogen atom and (b) the energy of the first orbit in the singly ionized helium atom He$^+$.

(a) The energy of the first orbit in hydrogen ($Z = 1$) is given by Equ. 3.4 with $n = 1$ so that $E_1 = -Rhc$ and

$$E_1 = -1 \cdot 097 \times 10^7 \text{ m}^{-1} \times 6 \cdot 63 \times 10^{-34} \text{ J s} \times 3 \cdot 00 \times 10^8 \text{ m s}^{-1}$$

$$E_1 = -2 \cdot 19 \times 10^{-18} \text{ J}$$

A more convenient unit of energy is the electronvolt (eV) where $1 \text{ eV} = 1 \cdot 60 \times 10^{-19}$ J. Using this unit $E_1 = -2 \cdot 19 \times 10^{-18} \text{ J}/1 \cdot 60 \times 10^{-19} \text{ J eV}^{-1} = -13 \cdot 6 \text{ eV}$. This energy is the ionization energy of the hydrogen atom and is the amount of energy which must be supplied to the atomic electron to remove it from the atom. The radius of the first orbit is given by Eq. 3.5

$$r_1 = \frac{-11 \cdot 5 \times 10^{-29} \text{ N m}^2 \times 1}{-2 \cdot 19 \times 10^{-18} \text{ J}}$$

$$r_1 = 5 \cdot 3 \times 10^{-11} \text{ m}$$

From Equ. 3.4 the energy of the second orbit $E_2 = E_1/(2)^2 = -3 \cdot 4 \text{ eV}$ and the radius $r_2 = r_1 \times (2)^2 = 2 \cdot 1 \times 10^{-10}$ m.

(b) For singly ionized helium (He$^+$) which has $Z = 2$, Eq. 3.4 gives $(E_1)_{He} = E_1 \times (2)^2 = -54 \cdot 4 \text{ eV}$ and $(r_1)_{He} = r_1 \times (2/4) = 2 \cdot 6 \times 10^{-11}$ m.

The allowed electron energy states in the hydrogen atom and the corresponding radii of the orbits are summarized in Fig. 3.3. This illustrates the discreteness in the energy levels available to the electron, the allowed states being separated by energies which are unstable and forbidden. When the electron is in the state corresponding to $n = 1$, the hydrogen atom is said to be in its ground state and the electron is most tightly bound in the atom. For n greater than 1, the atom is in an excited state and when $n = \infty$ (where $E_n = 0$) the electron is no longer bound in the atom and the atom is ionized. Once outside the atom, there is a continuous range of kinetic energies available to the electron and the restrictions of quantum theory no longer apply.

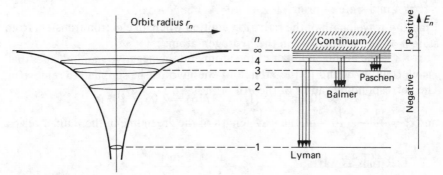

Figure 3.3 Energies and radii of allowed electron orbits in hydrogen, showing the principal emission series.

Any electronic transition from one stationary state to another involves a gain or loss of energy equal to the energy difference between the states, which may be absorbed or emitted as radiation. Fig. 3.4 shows the electron transition between two such states of energies E_1 and E_2, the photon absorbed or emitted having energy

$$E = hf = E_2 - E_1 \tag{3.6}$$

If an electron moves from an excited state where $n = n_2$ to a lower energy state where $n = n_1$ (i.e. n_2 greater than n_1), the energy of the radiation emitted is, from equations 3.4 and 3.6 with $Z = 1$ for hydrogen,

$$E = hf = -Rhc \left(\frac{1}{n_2^2} - \frac{1}{n_1^2} \right)$$

or, since $c = \lambda f$

$$\frac{1}{\lambda} = R \left(\frac{1}{n_1^2} - \frac{1}{n_2^2} \right) \tag{3.7}$$

in agreement with Equ. 3.2 for the Balmer series for which $n_1 = 2$ and $n_2 = 3, 4, 5$ etc. In this series, all transitions end in the state $n = 2$. Radiation is emitted in other series, the more common ones corresponding to transitions to state $n = 1$ (Lyman) and $n = 3$ (Paschen). The energies of the radiation emitted in the Lyman series are larger than for the Balmer series, the wavelengths being shorter and the radiation is ultraviolet light which is not visible. Similarly the lower energies of the Paschen series result in the longer wavelengths of infrared radiation. Transitions in these series are shown in Fig. 3.3.

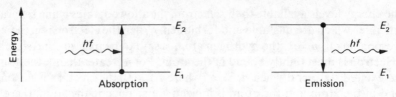

Figure 3.4 Electron transfer in radiant absorption and emission.

Example 3.3

Find the wavelength of the radiation emitted when an electron transfers from state $n = 4$ to state $n = 2$ in the hydrogen atom.

Using Equ. 3.7 with $n_1 = 2$ and $n_2 = 4$ we have

$$1/\lambda = 1\cdot097 \times 10^7 \, \text{m}^{-1} \times (1/4 - 1/16) = 2\cdot06 \times 10^6 \, \text{m}^{-1}$$

and $\lambda = 4\cdot86 \times 10^{-7}$ m, the wavelength of the green line in the Balmer series.

3.2 Electrons as waves

In the atomic context, light behaves like a particle, the photon, which has a definite energy. This view is opposed to our experience of light having a wave-like nature, as exhibited in phenomena such as interference and diffraction (Chapter 7). This apparent contradiction is due to the atomic system being so small that we are unable to envisage its properties on a large scale. This double character is called the wave–particle duality of light. In some cases light may be considered to behave like a wave, in others like a particle.

It is normal to consider the electron as a particle, the mass of which may be measured. However, in some instances, electrons (and other particles) behave as waves, with a wavelength given by

$$\lambda = h/Mv \tag{3.8}$$

where Mv is the momentum of the particle. The faster a particle moves, the shorter is its wavelength. This wave-like nature of electrons is utilized in the electron microscope (see Chapter 8).

Example 3.4

What is the wavelength associated with an electron ($M = 9\cdot11 \times 10^{-31}$ kg) moving with speed $1\cdot0 \times 10^7$ m s^{-1}?

From Equ. 3.8, $\lambda = h/Mv$, we have

$$\lambda = \frac{6\cdot63 \times 10^{-34} \, \text{J s}}{9\cdot11 \times 10^{-31} \, \text{kg} \times 1\cdot0 \times 10^7 \, \text{m s}^{-1}} = 7\cdot3 \times 10^{-10} \, \text{m}$$

This new field of wave mechanics has greatly assisted our description of the atom. An orbiting electron may be considered as a wave, the amplitude of which is related to the probability of finding the electron in a particular position at some time. The atomic nucleus is surrounded by an electron cloud (a break

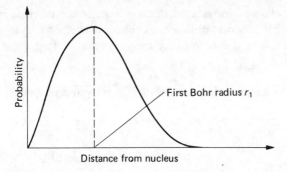

Figure 3.5 The probability distribution for a ground-state electron in hydrogen.

from the solar analogy in which the sun is surrounded by well defined planets), and the permitted electron energies, which previously corresponded to allowed orbits, are now associated with permitted waves. Fig. 3.5 shows the probability of finding a ground-state electron in the hydrogen atom at a distance r from the nucleus, derived from a wave mechanical treatment. The maximum probability occurs at the first Bohr radius r_1.

3.3 The electron shell structure of complex atoms

Although Bohr theory is useful for predicting the electron energy states of the hydrogen atom and the other single-electron atoms He^+ and Li^{2+}, a similar treatment for many-electron atoms is unsatisfactory since only the electrostatic attraction between the positively charged nucleus and the negatively charged electron is considered, and no account taken of the repulsion between electrons. This repulsion explains why the measured value of the ionization energy of the He atom is approximately one-half that calculated in Example 3.2. Also, in general the electron does not move around the nucleus in a spherical orbit. Wave mechanics shows that for a particular value of principal quantum number n the electron can have different quantized amounts of angular momentum. This is reflected in the shape of its orbit. Historically, these shapes have been labelled as s-orbitals which are spherical, p-orbitals which have two lobes, d-orbitals having several lobes, and others yet more complex. Some configurations are shown in Fig. 3.6. An electron in a p-orbital corresponding to $n = 3$ is designated a 3p-electron.

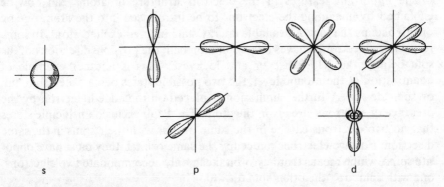

s   p   d

Figure 3.6 Some orbital shapes.

The number of orbitals allowed for any particular shape is summarized in Table 3.1. It can be seen that the total number of orbitals corresponding to a particular value of n, that is to a particular electron energy, is n^2. In practice

Table 3.1 Dependence of the number of atomic orbitals on the principal quantum number.

n	s	p	d	f	Total
1	1				1
2	1	3			4
3	1	3	5		9
4	1	3	5	7	16

however, the energies in the various orbitals are not identical and a 2p-electron is in a slightly higher energy state than a 2s-electron. Similarly, the 4f-state is slightly higher than the 4d-state, which itself is higher than either the 4p- or the 4s-states. A generalized energy level diagram is shown in Fig. 3.7.

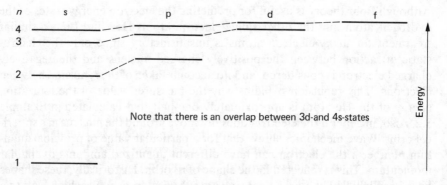

Note that there is an overlap between 3d-and 4s-states

Figure 3.7 The dependence of electron energy on orbital shape.

The important features of the electron structure of atoms can now be explained by imagining the electrons to be introduced into the atom one by one to occupy the lowest available energy state in an unexcited atom. In some respects an electron behaves like a spinning ball, the possible directions of the spin being either clockwise or anticlockwise. This is another example of quantization at the atomic level, the two possible states being called spin 'up' or spin 'down'. A further limitation is placed on the added electron by the presence of electrons already in the atom. The Pauli exclusion principle states that no two electrons can be in the same orbital with their spins in the same direction. For two electrons to occupy the same orbital, they must have opposite spins, which means that any orbital can only accommodate two electrons, one with spin 'up', the other spin 'down'.

The first electron introduced will occupy the state of lowest energy, the 1s-state. This is the situation in the hydrogen atom (Fig. 3.8). A second electron will try to occupy the same state, and can do so by aligning its spin antiparallel to the spin of the first electron (the helium atom). A third electron is unable to enter the 1s-state since this is fully occupied, and goes into the next lowest state, the 2s-state (the lithium atom). This procedure may be extended to elements of increasing atomic number. Electrons in states corresponding to $n = 1$ are termed K-shell electrons, those with $n = 2$ are L-shell electrons and those with $n = 3$ are M-shell electrons. In this way the shell structure of atoms emerges. Since the total number of orbitals in a shell of principal quantum number n is n^2 (see Table 3.1) and each orbital can accommodate two electrons, the total electron number in each shell is $2n^2$. Thus the K-shell contains 2 electrons ($2 \times 1s$), the L-shell 8 electrons ($2 \times 2s$, $6 \times 2p$), the M-shell 18 electrons ($2 \times 3s$, $6 \times 3p$, $10 \times 3d$) etc.

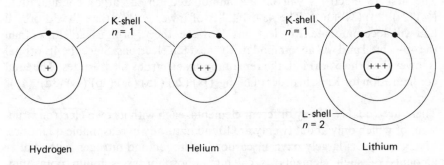

K-shell
$n = 1$

K-shell
$n = 1$

L-shell
$n = 2$

Hydrogen   Helium   Lithium

Figure 3.8 Simple atomic shell structure.

Atoms having a full electron complement in their shells are more stable, or less chemically reactive, than atoms with either a single electron in an otherwise empty shell or a single vacancy in an otherwise full shell. If the number of electrons in an orbital is denoted by a superscript (for example, hydrogen has a configuration $(1s)^1$), then on this basis neon ($Z = 10$; $(1s)^2(2s)^2(2p)^6$) is very stable as the K- and L-shells are filled and the M-shell is empty. Conversely, fluorine ($Z = 9$; $(1s)^2(2s)^2(2p)^5$) having a vacancy in the L-shell and sodium ($Z = 11$; $(1s)^2(2s)^2(2p)^6(3s)^1$) with a single electron in the M-shell must form chemical bonds with other atoms to achieve stable electron configurations. This will be discussed later.

Unfortunately, the electron structures of atoms cannot always be explained on such a simple basis. We have seen that, in a shell characterized by principal quantum number n, the p-electrons are at a slightly higher energy level than the s-electrons, and the d-electrons slightly higher than the p-electrons. As n increases, the d-electrons in a particular shell tend to have a higher energy than the s-electrons in the next shell, as shown in Fig. 3.7 when $n = 3$. Thus the 4s-states are filled before the 3d-states and the 6s-states before the 4f-states. The order in which the electron sub-shells are filled as the atomic number increases is

1s, 2s, 2p, 3s, 3p, 4s, 3d, 4p, 5s, 4d, 5p, 6s. . .

The electronic configuration of potassium ($Z = 19$) is $(1s)^2(2s)^2(2p)^6(3s)^2(3p)^6$ $(3d)^0(4s)^1$. The single electron in the N-shell explains why potassium is chemically similar to sodium, which has a single electron in the M-shell. Argon ($Z = 18$; $(1s)^2(2s)^2(2p)^6(3s)^2(3p)^6$) has filled sub-shells and is chemically unreactive. Due to these energy shifts in the outer shells, the chemical reactivity of the elements depends on sub-shell structure, closed sub-shells occurring when $Z = 2$(He), 10(Ne), 18(Ar), 36(Kr), 54(Xe) and 86(Rn). These are the inert gases. This may be contrasted with the closed-shell situation based on $2n^2$ electrons per shell, which predicts inert gases when $Z = 2, 10, 28$(Ni) and 60(Nd).

Example 3.5

Show that the chemical stability of krypton ($Z = 36$) may be explained on the basis of completely filled sub-shells.

The first 18 electrons will be accommodated on the argon configuration: $(1s)^2(2s)^2(2p)^6(3s)^2(3p)^6$. The 4s-orbital is of lower energy than the 3d-orbital and will accommodate two electrons. However the 3d-orbital is still lower than the 4p-orbital and will accommodate the next ten electrons. Since the 4p-orbital is lower than the 5s-orbital, the remaining six electrons fill 4p-states. The total configuration for Kr is, in order of filling, $(1s)^2(2s)^2(2p)^6(3s)^2(3p)^6(4s)^2(3d)^{10}(4p)^6$.

There are roughly ninety different elements, each with its own electron structure, of which only about twenty are found naturally in reasonable quantities. Living matter is largely oxygen, carbon, hydrogen and nitrogen with smaller amounts of such elements as calcium, phosphorous, sulphur, potassium, sodium and chlorine. Table 3.2 shows the percentage by weight of the main body elements. Elements such as silicon, iron, aluminium and magnesium are in abundance in the earth's crust, and the million or so different substances presently classified arise largely from the way these relatively few elements combine.

Table 3.2   The major body elements.

Element	% Weight	Element	% Weight	Element	% Weight
Oxygen	65	Calcium	1·5	Sodium	0·15
Carbon	18	Phosphorus	1	Chlorine	0·15
Hydrogen	10	Sulphur	0·25	Magnesium	0·05
Nitrogen	3	Potassium	0·2	Iron	0·006

3.4   Molecular structure

The chemical properties of the elements differ because of the way orbital electrons interact to form interatomic bonds. A molecule is formed whenever a system of atoms bonded together is more stable than the isolated atoms. The only group of elements whose atoms normally exist separately are the inert gases. Helium, neon and argon form no chemical compounds, and krypton, xenon and radon form a few in combination with fluorine and oxygen. Their reluctance to form chemical bonds is due to the outer electron shell being completely filled with no vacancies to accommodate other electrons, and no lone electrons which may be donated to fill vacancies in the electronic shells of other atoms. Thus the ability to form interatomic bonds is related to the structure of the outermost, or valence, electron shells. Electrons in the inner subshells are termed core electrons and are tightly bound in the atom. They have little effect on its chemical properties.

Alkali metals such as lithium, sodium and potassium have a single valence electron, whereas the halogens (fluorine, chlorine, bromine and iodine) have an outer shell with a vacancy. An alkali halide molecule is formed when the valence electron is transferred from the alkali metal to fill the outer shell of the halogen atom, the two electrically neutral atoms forming a molecule consisting of two ions held together by an ionic bond. This is demonstrated by the reaction

$$Na + Cl \rightarrow Na^+ + Cl^- \tag{3.9}$$

in which an electron is transferred from the sodium atom to the chlorine atom. The sodium atom, having lost an electron, has a net positive charge (positive ion) and the chlorine atom with a surplus electron is negatively charged (negative ion). This is illustrated in Fig. 3.9a. In ionic bonding the cohesive force is the electrostatic attraction between oppositely charged ions (see § 1.4). This electrostatic attractive force extends over many atomic diameters and several ions may be attracted towards a single oppositely charged ion. It is called a long-range force. In crystalline NaCl for example each Na^+ ion is surrounded by six Cl^- ions as shown in Fig. 3.9b.

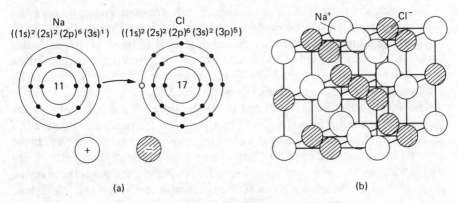

Figure 3.9 (a) The ionic bond. (b) Long-range order in crystalline NaCl.

If two atoms of the same element come together, charge is not transferred permanently from one to the other because the electronic structures of the two atoms are identical. Nevertheless, two hydrogen atoms are able to join together to form the very stable hydrogen molecule. Each atom has an electron in a shell which is able to accommodate two, and the orbital is only half-filled. Therefore hydrogen has a high attraction for electrons and tends to pull loosely attached electrons from other atoms. Where only hydrogen atoms are present, such electrons are not available, but as the two atoms come together, the half-filled orbitals of each atom overlap to form a new molecular orbital containing two electrons. Each electron loses its identity with the original atom and is shared equally. The new molecular orbital constitutes a covalent (or shared-electron) bond. Since one orbital is involved for each atom, this molecule is said to contain a single bond. Another example of covalent bonding is the oxygen molecule, atomic oxygen ($Z = 8$; $(1s)^2(2s)^2(2p)^4$) having two vacancies in the 2p-orbitals. The oxygen molecule is formed by two electrons from each atom being shared in molecular orbitals, forming a double covalent bond and effectively a closed outer shell. The covalent bonding of molecular hydrogen and oxygen is shown in Fig. 3.10. Covalent bonds can also occur between atoms of different elements. The HCl molecule is formed in this way, the hydrogen and chlorine atoms sharing electrons to achieve stable filled-shell configurations.

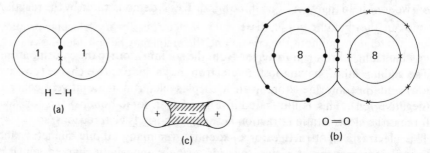

Figure 3.10 Covalent bonding in (a) a hydrogen molecule and (b) an oxygen molecule. (c) A simplified diagram of the charge distribution in these bonds.

The electrons participating in covalent bonds are contributed by both the bonded atoms and exist in molecular orbitals around them. The bond is localized, but each atom may be bonded to other atoms, using different electrons. In this way, the water molecule H_2O is formed by two covalent bonds, each linking a hydrogen atom to the oxygen atom. Similarly ammonia NH_3 and methane CH_4 are formed by covalent bonds as shown in Fig. 3.11.

The short-range, highly-localized cohesive force of covalent bonding explains the formation of molecules of relatively few atoms usually existing in a gaseous or liquid state, but larger molecules may be formed in this way. In the paraffin hydrocarbons, up to fifty carbon atoms are covalently bonded in chains with hydrogen atoms linked on either side. A simple example is the propane molecule C_3H_8 which has a chain of three carbon atoms, whereas polyethylene contains molecules having more than ten thousand carbon atoms.

A metallic solid is formed when atoms which have one or more valence electrons are held together in a regular or periodic crystal structure similar to that shown in Fig. 3.9b. In contrast to the localized covalent bonding in non-metals, metallic atoms such as sodium, copper and gold give up these valence electrons which are then free to move about within the crystal lattice structure. The metallic bond is formed because of the electrostatic attraction of the positive ions for the surrounding negatively charged electrons. It is impossible to determine which electron belongs to a particular atom. For example, a lithium

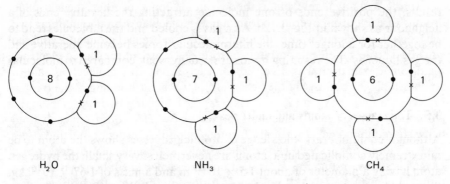

Figure 3.11   Covalent bonding in common polyatomic molecules.

atom has a single valence electron $((2s)^1)$, and this element may occur in gaseous form as diatomic Li_2, the valence electrons forming a single covalent bond. In the metallic solid, these valence electrons form molecular orbitals which extend throughout the crystal and the metal may be considered as a very large molecule comprising a skeleton of positively charged ions surrounded by a sea of electrons.

Some indication has been given of the mechanisms by which ionic- and metallic-bonded molecules interact to form solids and liquids, but it is difficult to understand the origin of the intermolecular cohesion of covalently bonded molecules. This arises from the weak, non-directional van der Waals' force which is responsible for holding certain solids and liquids together and is the basis of the action of adhesives. A single atom is electrically neutral and the positive charge on the nucleus is located, on average, at the centre of the negative electronic charges. If a second atom is brought close, the nucleus of the first atom is attracted by the electron cloud of the second whereas the electron clouds of the two atoms repel each other. The net result is a displacement of the nucleus from the centre of the atom, and the atom is said to be polarized. This situation is shown schematically in Fig. 3.12a. This polarizing effect also occurs in covalently bonded molecules of dissimilar atoms such as water, in which the bonding electrons spend more time in the vicinity of the oxygen atom than the hydrogen atoms, and the hydrogen atoms are effectively positive ions whereas the oxygen atom carries a net negative charge. A polarized situation

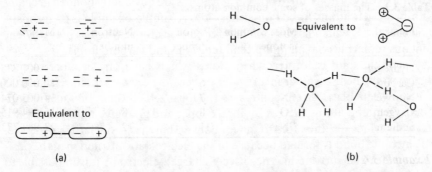

Figure 3.12   (a) Atomic and molecular polarization. (b) Intermolecular bonding in water.

results, the positive ends of one molecule attracting the negative ends of a neighbour as shown in Fig. 3.12b. As a gas is cooled and the molecules tend to be together for a longer time, the intermolecular forces become operative and the gas liquefies. The resulting bonds are usually weak compared to molecular bonds.

3.5   The masses of atoms and molecules

Although contemporary knowledge of atomic processes shows the atom to be an extremely complicated unit, atoms are nevertheless very small, the hydrogen atom having a diameter of about $1 \cdot 5 \times 10^{-10}$ m and a mass of $1 \cdot 67 \times 10^{-27}$ kg. Since such masses based on the kilogram are numerically difficult to manipulate, a more suitable mass scale has been devised for the atom.

Any element has an atomic number Z, equal to the number of protons in the nucleus and to the number of orbital electrons. However the nucleus contains another particle, the neutron, which has a similar mass to the proton but carries no charge. The atoms of a particular element may contain varying numbers of neutrons, and so have different masses. These different masses correspond to different isotopes of the elements. For example, the isotope oxygen-16 (denoted ^{16}O), having eight protons and eight neutrons has a larger mass than oxygen-15 (denoted ^{15}O), with eight protons but only seven neutrons.

The unit of mass now employed is the unified atomic mass unit (u), the isotope carbon-12 (^{12}C, six protons, six neutrons) being defined as having a mass of exactly 12 u. Although the SI unit of the amount of substance is the mole (mol) it is often more convenient to use the accepted alternative, which is the kilogram mole (kmol). Since 1 kmol of any substance contains N_A molecules, where N_A is the Avogadro constant, and 1 kmol of ^{12}C has a mass of 12 kg, then one atom has a mass of 12 kg/N_A. Therefore from the above definition $u = (1/12) \times (12 \text{ kg}/N_A)$ and since $N_A = 6 \cdot 022 \times 10^{26}$ molecules kmol^{-1},

$$u = 1 \text{ kg}/(6 \cdot 022 \times 10^{26}) = 1 \cdot 66 \times 10^{-27} \text{ kg}$$

On this atomic scale, the masses of the most common isotopes of several elements are shown in Table 3.3, illustrating the close relation between mass and number of protons and neutrons in the nucleus.

Table 3.3   The masses of some common atoms.

Element	Most common isotope	Proton number	Neutron number	Mass/u
Hydrogen	^1H	1	0	1·007 825
Carbon	^{12}C	6	6	12·000 000
Nitrogen	^{14}N	7	7	14·003 074
Oxygen	^{16}O	8	8	15·994 915
Sodium	^{23}Na	11	12	22·989 773

Example 3.6

Use Table 3.3 to calculate the mass in kg of one molecule of water.

From Table 3.3 the mass of one molecule of H_2O in atomic mass units is

$$M_u = 2 \times 1 \cdot 007\ 825\ u + 1 \times 15 \cdot 994\ 915\ u = 18 \cdot 010\ 565\ u$$

Since $1\ u = 1 \cdot 66 \times 10^{-27}$ kg, the mass of a water molecule is

$$M = 18 \cdot 010\ 565\ u \times 1 \cdot 66 \times 10^{-27}\ \text{kg u}^{-1} = 2 \cdot 989 \times 10^{-26}\ \text{kg}$$

The masses given in Table 3.3 have been determined by the technique of mass spectrometry, which analyses the atoms and molecules in a sample on the basis of mass. This analysis may be made using a Bainbridge mass spectrograph (shown in Fig. 3.13) which, although lacking modern sophistication and accuracy, nevertheless illustrates the principle of the technique.

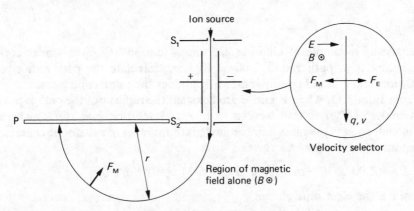

Figure 3.13   The Bainbridge mass spectrograph.

In the ion source, the sample to be analysed is vaporized and bombarded with energetic electrons which may remove one or more orbital electrons from the molecules and form positive ions which are then accelerated through slit S_1. Upon emerging from the ion source the ions have a range of speeds, charge and mass. Mass determination depends on all ions having the same speed, and this is achieved in the 'velocity selector', the region between S_1 and S_2. A uniform magnetic field is applied perpendicular to and out of, the plane of the diagram everywhere below S_1. In the velocity selector region, an electric field is maintained between the two plates as shown. A positive ion travelling through the velocity selector experiences a force to the right due to the electric field and to the left due to the magnetic field (see §1.4). Such an ion can only pass through S_2 if these forces are equal and it is not deflected from a straight path. From Equ. 1.18 the electric force $F_E = qE$ where q is the charge on the ion and E is the electric field, and the magnetic force (Equ. 1.20) is $F_M = Bqv$ where B is the magnetic flux density and v is the speed of the ion. The ions pass through the selector when $F_E = F_M$ or $qE = Bqv$ so that

$$v = E/B \tag{3.10}$$

The speed of the transmitted ions is changed by varying either E or B. For an electric field of $5 \cdot 0 \times 10^5$ V m^{-1} and a magnetic field of $0 \cdot 20$ T (tesla) the speed of the transmitted ions $v = 5 \cdot 0 \times 10^5/0 \cdot 20 = 2 \cdot 5 \times 10^6$ m s^{-1}. Once through the velocity selector, the ions enter a region where they are influenced by the magnetic field alone, and move in a circular path of radius r to hit a photographic plate P. This circular orbit is maintained by the centripetal force $F_c = Mv^2/r$ due to the magnetic field, which is directed towards the centre of the circle. From Equ. 1.21 the radius of the path is $r = Mv/Bq$, and substituting for v from Equ. 3.10 gives

$$r = (M/q)(E/B^2) \tag{3.11}$$

The radius of the path, as determined by the point of impact of the ion on the photographic plate, is proportional to (M/q) and the relative number of such ions in a sample is determined from the degree of blackening of the plate.

Example 3.7

In analysing a sample of chlorine gas, singly ionized ^{35}Cl gives a peak corresponding to a path radius $2 \cdot 00 \times 10^{-2}$ m. Calculate the path radius of singly ionized ^{37}Cl and doubly ionized ^{35}Cl under the same conditions.

From Equ. 3.11, when E and B are constant the radius r of the path is proportional to (M/q) so that we can write $r = k(M/q)$ or $k = r(q/M)$ where r is in metres, M is the mass number and q is the number of net atomic charges. Substituting values for $^{35}Cl^+$ gives

$$k = (2 \cdot 00 \times 10^{-2} \text{ m} \times 1 \text{ } e)/35 \text{ u} = 5 \cdot 71 \times 10^{-4} \text{ m e u}^{-1}$$

where e is the electronic charge.

For $^{37}Cl^+$, $r = (5 \cdot 71 \times 10^{-4} \text{ m e u}^{-1} \times 37 \text{ u})/1 \text{ } e = 2 \cdot 12 \times 10^{-2}$ m

and

for $^{35}CL^{2+}$, $r = (5 \cdot 71 \times 10^{-4} \text{ m e u}^{-1} \times 35 \text{ u})/2 \text{ } e = 1 \cdot 00 \times 10^{-2}$ m.

A typical mass spectrum for chlorine gas is shown in Fig. 3.14, illustrating the

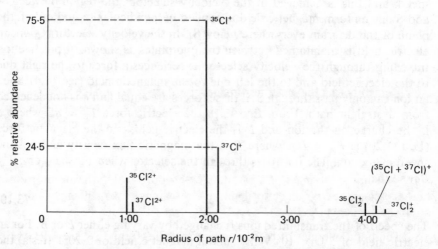

Figure 3.14 The distribution and relative abundance of ions from a chlorine gas sample.

relative abundances of ^{35}Cl and ^{37}Cl and indicating the existence and relative positions of the three possible Cl_2 molecules.

The more sophisticated instruments are able to measure masses extremely accurately and may determine not only the molecular weights but also the chemical formulae of molecules. For example, an instrument of moderate mass resolution can differentiate between a mixture of CO, N_2 and C_2H_4, each having molecular weight 28. Mass spectrometers of direct application in biological measurement are increasingly used since the recent growth in instrument technology has improved portability and simplicity of operation. Spectrometers used in lung volume determinations or to measure the amount of oxygen in the blood (an indication of metabolic activity) require a mass resolution of about one part in one hundred.

Problems

3.1   Excited iron atoms emit green light of wavelength 527 nm. Calculate the energy of these photons and compare the value with that of photons from BBC Radio 2 of wavelength 1500 m.

3.2   The wavelengths of the three visible lines in the Balmer series for hydrogen are 656·3 nm (red), 486·1 nm (green) and 435·8 nm (blue). By reference to the equation relating λ to n for this series, plot a suitable graph to find the value of the constant R.

3.3   The angular momentum of an electron in the nth orbit of an atom is $nh/2\pi$. Using the value of orbit radius calculated in Example 3.2, determine: (a) the angular momentum, (b) the linear momentum and (c) the speed of an electron in the K-shell of the hydrogen atom if the mass of the electron is $9\cdot11 \times 10^{-31}$ kg.

3.4   Calculate the energy, in eV, required to ionize the hydrogen atom from state $n = 5$. If this energy is provided by a photon, what is the minimum frequency of the photon which will produce such an ionization?

3.5   What is the wavelength associated with: (a) a sprinter ($M = 70$ kg) running at 10 m s^{-1}; (b) a snail ($M = 5\cdot0 \times 10^{-3}$ kg) moving at 100 mm per day; (c) an electron in the K-shell of the hydrogen atom (see Problem 3.3); and (d) the earth ($M = 6\cdot0 \times 10^{24}$ kg) moving around the sun at 30 km s^{-1}?

3.6   Remembering the order in which orbitals are filled and the occurrence of overlapping energy levels, determine the electron configurations of the halogen iodine ($Z = 53$), the inert gas xenon ($Z = 54$) and the alkali metal caesium ($Z = 55$).

3.7   Table 3.2 shows the percentage by weight of the main elements in the body. Assuming the mass of an atom to be proportional to the atomic weight of that element, derive an equivalent table showing the percentage by number of all the elements listed.

3.8   The bond energies of the diatomic molecules 0_2 and N_2 are $4\cdot93 \times 10^8$ J kmol^{-1} and $9\cdot40 \times 10^8$ J kmol^{-1} respectively. What is meant by bond energy, and what are the reasons for the difference in these values? Calculate the energy, in eV, required to separate each molecule into its constituent atoms. The hydrogen bond between water molecules has energy $5 \cdot 00 \times 10^7 \, \text{J kmol}^{-1}$. Give reasons for its low value compared with those above. What frequency of radiation is required to dissociate two water molecules?

3.9   Draw diagrams to show the electronic configurations in the ionic bonding of LiF and the covalent bonding of Cl_2.

3.10   Use Table 3.3 to calculate the molecular masses, in kg, of ammonia (NH_4) and glycerol $(C_3H_8O_3)$.

3.11   A Bainbridge mass spectrograph is used with $E = 1 \cdot 0 \times 10^4 \, \text{V m}^{-1}$ and $B = 0 \cdot 20$ T. Calculate the speed of ions passing through the velocity selector. If the ion source produces singly charged ions of the carbon isotopes ^{12}C and ^{13}C and the photographic plate lies along the diameter of the circular path travelled by the ions in the magnetic field, find the distance between the two images produced on the plate. Assume the mass of an atom of ^{13}C to be 13 u.

3.12   Use values of atomic masses given in Table 3.3 to calculate the masses of CO, N_2 and C_2H_4 molecules. Use a suitable scale to show the mass spectrum expected from a mixture of 25% CO, 45% N_2 and 30% C_2H_4.

4

Spectroscopy

The atoms and molecules which constitute all materials are very small, making microscopic examination impossible. However, as has been shown, the energy levels of such molecules are discrete, or quantized, and any transition to a higher or lower energy level involves the absorption or emission of a well defined amount of energy, usually in the form of a photon. By measuring the energy of the radiation absorbed or emitted it is possible to analyse a sample into its molecular components and, further, by determining the amount of such radiation involved, we can estimate the quantity of a particular molecular species present. The field of spectroscopy is concerned with the interaction of electromagnetic radiation with matter.

4.1 The electromagnetic spectrum

As was shown in the last chapter, excited hydrogen atoms emit light on returning to the ground state, the three lines visible being the emission spectrum of hydrogen in that range of wavelengths which the eye can detect. This visible region, which extends from approximately $\lambda = 400$ nm to 700 nm, is a small part of the complete electromagnetic spectrum which ranges from radio waves (λ about 10^3 m) to gamma rays (λ about 10^{-12} m). Gamma rays arise from radioactive nuclear decay and are discussed in Chapter 11. The full electromagnetic spectrum is shown in Fig. 4.1, and includes the equivalent frequency ranges (obtained from $f = c/\lambda$) and the corresponding photon energies E. The divisions between neighbouring regions of the spectrum are not precise, the classification arising more from the origin of the radiation than its wavelength and overlapping occurs because of this.

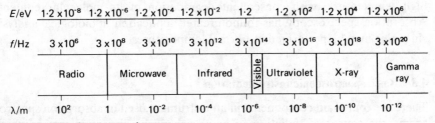

Figure 4.1 The electromagnetic spectrum.

Example 4.1

An electromagnetic wave has photons of energy $2 \cdot 00 \times 10^{-17}$ J. Express this energy in electronvolts and classify the wave in the electromagnetic spectrum.

Since 1 eV = $1 \cdot 60 \times 10^{-19}$ J the energy of the wave is

$$E = \frac{2 \cdot 00 \times 10^{-17} \text{ J}}{1 \cdot 60 \times 10^{-19} \text{ J eV}^{-1}} = 1 \cdot 25 \times 10^2 \text{ eV}$$

By referring to Fig. 4.1 it can be seen that this is an ultraviolet wave.

4.2 The absorption and emission of radiation

A material sample contains atoms and molecules in different physical and chemical forms. Although such a sample may appear completely motionless this is by no means the situation at the molecular level. Molecules, such as those in a gas are able to move through the material system and exhibit translational motion. They may also rotate about some molecular axis, or a bond linking the atoms may vibrate, similar to the vibrational motion of a mass on a spring. Additionally, electrons in the molecule may change energy states. In general a molecule will have all four components to its total motion. Fortunately we are able to analyse each component separately.

All motion is concerned with energy. Thus a rotating molecule has rotational energy and a vibrating molecule has vibrational energy. If its vibrational energy increases then it vibrates with larger amplitude and changes to a state of higher vibrational energy. However, in rotational and vibrational motion, the energy states available to the molecule are not continuous and these motions occur at discrete energies, just as atomic electron energy states are quantized. For this reason a molecule can only rotate and vibrate at certain well defined frequencies which are specific to that molecule. For example, a molecule may absorb radiation to increase its vibrational energy by transferring from a state of vibrational energy E_1 to a higher energy state E_2, the energy hf of the photon absorbed being equal to the energy difference between the two states, $hf = E_2 - E_1$. Photons of frequency larger or smaller than this cannot be absorbed. This situation also applies in photon emission during transfer from a higher to a lower energy state. Because of the specific nature of the frequencies at which a molecule will absorb or emit radiation, measurement of such frequencies enables the molecular composition of the sample to be determined. This is the basis of spectroscopy.

The emission frequencies are usually measured with the sample at a relatively high temperature, with consequent degradation of its molecular content. Absorption spectroscopy has found greater application in biological analysis since this may be performed at normal temperatures.

4.3 Basic spectroscopic instrumentation

There are four main components in an instrument used in absorption spectroscopy; the source of radiation to be absorbed, the dispersion system which splits this radiation into its component wavelengths, the sample to be analysed and the detector which measures the degree of absorption in the sample.

Any source of radiation used should provide radiation over the wavelength range of interest and also have an intensity distribution which varies smoothly over this range with no sharp discontinuities. The sun is a copious source of visible and infrared radiation and if the 'white' light from the sun is split into its component wavelengths and the number of photons arriving each second at each wavelength is counted, an emission spectrum is obtained showing the variation in intensity with wavelength (Fig. 4.2). Although the intensity of the sun's radiation is not constant over the visible region, there are no sharp discontinuities in the spectrum. Of course, this source of radiation is not used in the laboratory because of fluctuations in intensity and because oxygen, ozone, water vapour and carbon dioxide in the atmosphere absorb sunlight at specific wavelengths so producing discontinuities. However the theme of sun as source and eye as detector will be developed further to illustrate the principle of spectroscopy.

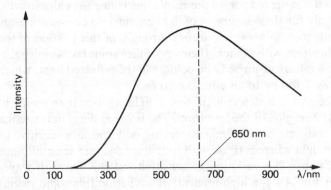

Figure 4.2   The emission spectrum of the sun.

The separation of sunlight into its component wavelengths occurs in the rainbow, where small water droplets in the atmosphere deviate, or refract, different wavelengths through different angles (see § 7.1) and effectively produce the sunlight spectrum consisting of the colours of the rainbow: red, orange, yellow, green, blue, indigo and violet (in order of decreasing wavelength). The same spectrum may be produced using a glass prism (Fig. 4.3a), which deviates the shorter wavelengths through larger angles than the longer wavelengths. In this way violet light is deviated through a larger angle than red light and appears at the 'bottom' of the spectrum. The prism is an example of a dispersion system employed in some mono-chromators. Two other types of monochromator commonly used in spectro-scopy are the diffraction grating and the filter. The diffraction grating consists of an array of parallel, closely spaced lines ruled on a glass plate. This sepa-rates a beam of radiation into its constituent wavelengths by the diffraction of light (see § 7.6). The filter transmits only a small wavelength range of the radiation incident on it, and absorbs all others. The function of a filter is illustrated by imagining a piece of yellow glass to be placed in the beam of white light as shown in Fig. 4.3b. Subsequent analysis of the spectrum shows that only yellow light is transmitted through the system, because components in the yellow-coloured glass absorb radiation at all frequencies except the yellow frequencies. That is, the energy states present are such that reds, greens and blues are absorbed but not yellows.

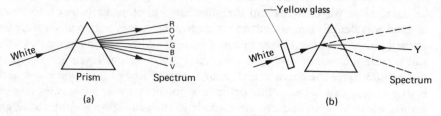

Figure 4.3   (a) Dispersion by a prism. (b) The effect of a filter.

The green colour of foliage may be explained using this subtractive argument. Fig. 4.2 shows that the peak in the emission spectrum of the sun occurs at about 650 nm, which is in the red part of the visible spectrum. The chlorophylls, which are responsible for the conversion of light energy into chemical energy (photosynthesis) are molecules which absorb strongly in this portion of the spectrum (and so function with higher efficiency). Since some red wavelengths of the incident sunlight are absorbed, the colours in the reflected light, which is detected by the eye, add up to an effective green.

The retina of the eye is a detector of visible light, that is, it responds to radiation in the visible region of the spectrum. The nerve endings in the retina are of two types; rods and cones. Animals having only rods are sensitive to the intensity of the light whereas those with rods and cones are sensitive both to intensity and colour. The variation of sensitivity of the eye with wavelength is shown in Fig. 4.4 for high intensity illumination (photopic vision) and low intensity illumination (scotopic vision). In broad daylight the eye is most sensitive to yellow-green light (λ approximately 550 nm). At dusk the maximum sensitivity is at a shorter wavelength, about 500 nm, which is a green colour. In dim light the cones cease to function and the sensitivity of the eye is due to the rods alone, and all objects appear in various shades of

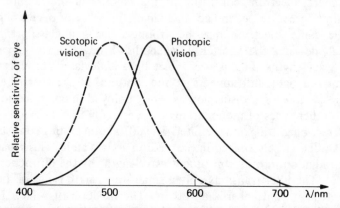

Figure 4.4   The wavelength response of the eye.

grey. Since the eye is most sensitive to light from 'green' objects under these conditions, leaves on trees frequently appear a lighter shade of grey than 'red' or 'blue' flowers. It must be remembered that the eye does not have its own monochromator, and that colour is not necessarily synonymous with wavelength. For instance, when red and blue light are added together to give a purple light, the eye 'sees' purple light rather than the two components.

The simple natural system described enables us to determine absorbed wavelengths only on the basis of colour, and even so we require prior knowledge of the emission properties of the source. Thus if the yellow glass is placed in white light we know that other colours are absorbed because they are not visible and because we know they were initially present. This is an example of a crude spectroscope, in which wavelengths absorbed are not accurately measured and the determination is very qualitative. This is of little use in analysis of biological samples where an accurate determination of absorbed wavelengths is necessary to determine molecular constituents and where some comparison with the absorption produced in a standard of known molecular concentration may be required to assess the sample concentration. This type of measurement is the function of the spectrophotometer.

There are two optical systems employed in such instruments. In the single-beam instrument, radiation from the source passes through the sample, is dispersed and then detected, the intensity of radiation transmitted at each wavelength being recorded (see § 6.5). This is then compared to the radiation transmitted by a standard containing a known concentration of the substance to be determined. A block diagram of this apparatus is shown in Fig. 4.5a. The main disadvantage of single-beam optics is that measurements on sample and standard are not simultaneous, and variations of source emission or detector sensitivity with time may be important. This disadvantage is overcome in the double-beam instrument (Fig. 4.5b), which has two identical beams of radiation taken from the same source, one of which passes through the standard and the other through the sample to be analysed. In practice a chopper allows light to pass alternately through sample and standard, and the instrument measures the ratio of the intensity of the sample beam to the standard beam.

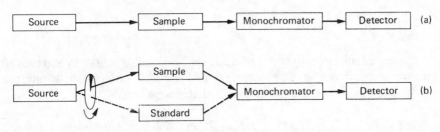

Figure 4.5 Spectroscopic instrumentation: (a) single-beam optics; (b) double-beam optics.

4.4 Beer–Lambert law of absorption

Whereas wavelengths at which absorption occurs determine which molecules are present in a sample, the degree of absorption allows estimation of the amounts present. If a beam of monochromatic radiation (radiation of a single wavelength) of intensity I_0 photons per unit area per second is incident on some sample and the intensity transmitted is I as shown in Fig 4.6a, then the absorbance (or optical density) is defined as

$$A = \log_{10}(I_0/I) \tag{4.1}$$

If a particular sample absorbs 50% of the incident beam, $I_0/I = 2$ and the absorbance $A = \log_{10}(2) = 0\cdot301$.

The absorbance depends on the number of molecules present in the sample. If the molecular concentration is c and the sample length is L then $A \propto cL$ or

$$A = acL \tag{4.2}$$

where a is the absorptivity of the sample at the wavelength considered. If the concentration is measured in mol m^{-3} and the thickness in metres, then, since A is a dimensionless number the units of a are m^2 mol^{-1}. Substituting for A from Equ. 4.1 we have

$$\log_{10}(I_0/I) = acL \tag{4.3}$$

which is the mathematical expression of the Beer–Lambert law of absorption relating absorbance to molecular concentration and sample length.

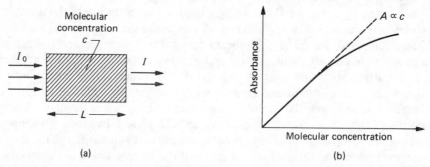

Figure 4.6   (a) The absorption of radiation in a sample. (b) Deviation from the Beer–Lambert law at high concentrations.

Example 4.2

An aqueous solution of yeast nucleic acid of length $2\cdot0 \times 10^{-2}$ m and concentration $2\cdot0$ mol m^{-3} has an absorptivity of 25 m^2 mol^{-1} at a certain wavelength. Calculate the percentage transmission at this wavelength.

From Equ. 4.3, $\log_{10}(I_0/I) = acL = 25$ m^2 mol^{-1} $\times 2\cdot0$ mol m^{-3} $\times 2\cdot0 \times 10^{-2}$m $= 1\cdot0$ so that $(I_0/I) = 10$ and the transmission is

$$I/I_0 = 0\cdot1 = 10\%$$

Equ. 4.2 shows that the absorbance increases with sample length. A more relevant observation is that in cases where L is determined by the dimensions of the sample holder the absorbance depends only on concentration. By measuring the absorbance of a standard solution of known concentration, the concentration of any sample may be determined by comparison. For example, if a standard of concentration c_0 has an absorbance A_0, a measured absorbance A in a sample of equal length corresponds to a concentration $c = c_0 A/A_0$. This method of calculating solution concentrations from absorbance measurements is widely used in spectroscopy. The Beer–Lambert law applies exactly only under optimized conditions and at low concentrations. At higher concentrations the relationship between A and c becomes non-linear as shown in Fig. 4.6b. This difficulty is overcome by preparing a calibration curve using solutions of known concentrations. By plotting the variation of absorbance with concentration, the concentration of any solution may be determined by measuring its absorbance.

4.5 Molecular rotation

Molecules in a gas may absorb certain frequencies of radiation so that they rotate faster. The kinetic energy of a particle of mass M moving with velocity v in a circular path of radius r is $E = Mv^2/2 = Mr^2\omega^2/2$ where ω is the angular velocity of the particle. Fig. 4.7a shows a diatomic molecule, with an atom of mass M_1 a distance r_1 from the axis of rotation and an atom of mass M_2 a distance r_2 from the axis. The molecular diameter is $r = r_1 + r_2$. The total kinetic energy of the molecule is $E = (M_1 r_1^2 \omega^2 + M_2 r_2^2 \omega^2)/2$, which may be written as

$$E = I\omega^2/2$$

where I is the moment of inertia of the diatomic molecule, equal to $(M_1 r_1^2 + M_2 r_2^2)$. It can be shown that

$$I = \mu r^2 \tag{4.4}$$

where μ is the reduced mass of the system given by

$$\mu = M_1 M_2/(M_1 + M_2) \tag{4.5}$$

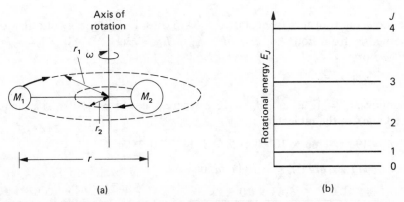

(a)   (b)

Figure 4.7   (a) The rotational motion of a diatomic molecule. (b) The rotational energy levels.

Example 4.3

Calculate the moment of inertia of the NO molecule if the bond length is 0·115 nm and the masses of nitrogen and oxygen atoms are $2·32 \times 10^{-26}$ kg and $2·66 \times 10^{-26}$ kg respectively.

The reduced mass of the system is, from Equ. 4.5,

$$\mu = (2·32 \times 10^{-26} \text{ kg} \times 2·66 \times 10^{-26} \text{kg})/(4·98 \times 10^{-26} \text{kg})$$

$$\mu = 1.24 \times 10^{-26} \text{ kg}$$

Substituting in Equ. 4.4 we have

$$I = 1·24 \times 10^{-26} \text{kg} \times (1·15 \times 10^{-10} \text{m})^2$$

or

$$I = 1·64 \times 10^{-46} \text{ kg m}^2$$

The rotational energy states available are not continuous but again are quantized, the energies of the allowed states being given by

$$E_J = h^2 J(J + 1)/(8\pi^2 \mu r^2) \tag{4.6}$$

where J is the rotational quantum number taking values 0, 1, 2, 3 etc. A schematic diagram of these rotational energy states is shown in Fig. 4.7b. A molecule which absorbs radiation transfers to a higher energy state and rotates faster. The energies involved are very small, corresponding to wavelengths in the microwave region of the electromagnetic spectrum (λ approximately 1 m to 1×10^{-3} m). Microwave absorption spectroscopy is still relatively unexploited in molecular analysis due to experimental difficulties of microwave production and detection.

Example 4.4

A CO molecule in the lowest rotational energy state absorbs radiation of frequency $1·16 \times 10^{11}$ Hz in transferring to the first excited state. Calculate the bond length of this molecule.

The molecule transfers from state $J = 0$ to $J = 1$. The energy of radiation absorbed is, from Equ. 4.6, $E = hf = E_1 - E_0 = h^2(1 \times 2 - 0 \times 1)/8\pi^2 I = 2h^2/8\pi^2 I$ or

$$I = h/4\pi^2 f$$

Assuming $M_C = 12$ u and $M_O = 16$ u where $1 \text{ u} = 1·66 \times 10^{-27}$ kg the reduced mass of the molecule is

$$\mu = 192 \times 1·66 \times 10^{-27} \text{kg}/28 = 1·14 \times 10^{-26} \text{ kg}$$

Since $I = \mu r^2 = h/4\pi^2 f$, $r^2 = h/4\pi^2 f\mu$, or

$$r^2 = \frac{6·63 \times 10^{-34} \text{J s}}{4\pi^2 \times 1·16 \times 10^{11} \text{s}^{-1} \times 1·14 \times 10^{-26} \text{kg}}$$

$$r^2 = 1 \cdot 27 \times 10^{-20} \text{ m}^2$$

or

$$r = 1 \cdot 13 \times 10^{-10} \text{ m} = 0 \cdot 113 \text{ nm}$$

4.6 Molecular vibration

The vibration frequency of a particle mass M suspended on a spring of force constant K is, from Equ. 2.7, $f_n = (1/2\pi)\sqrt{K/M}$. A diatomic molecule comprises two atomic masses, M_1 and M_2, connected by a chemical bond having force constant K, as shown in Fig. 4.8a. The molecule vibrates to and fro with vibration frequency

$$f_n = (1/2\pi)\sqrt{K/\mu} \tag{4.7}$$

where μ is the reduced mass given by Equ. 4.5. The vibrational energy states are again quantized such that

$$E_v = hf_n(v + \tfrac{1}{2}) \tag{4.8}$$

where v is the vibrational quantum number equal to 0, 1, 2, 3 etc. These equally spaced vibrational energy states are shown in Fig. 4.8b.

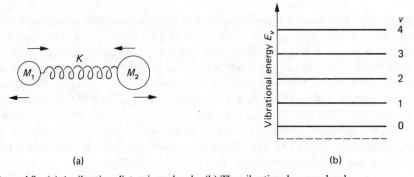

(a)   (b)

Figure 4.8 (a) A vibrating diatomic molecule. (b) The vibrational energy levels.

Example 4.5

The mechanical analogue of the CH_4 molecule is shown in Fig. 4.9. If the force constant of the $C-H$ bond is 500 N m^{-1}, find the frequency of radiation absorbed when the molecule transfers from one vibrational level to the one above.

The energy of the radiation absorbed is equal to the difference in energies of the two vibrational states, $E = hf = E_{v+1} - E_v$. From Equ. 4.8 $hf = hf_n$ or $f = f_n$. Taking $M_H = 1$ u and $M_C = 12$ u, the reduced mass is from Equ. 4.5,

$$\mu = 12 \times 1 \cdot 66 \times 10^{-27} \text{kg}/13 = 1 \cdot 53 \times 10^{-27} \text{kg}$$

Substituting in Equ. 4.7 gives

$$f = (1/2\pi)\sqrt{500 \text{ N m}^{-1}/1\cdot53 \times 10^{-27}\text{kg}}$$

or

$$f = (1/2\pi)\sqrt{32\cdot6 \times 10^{28}\text{s}^{-2}} = 9\cdot1 \times 10^{13}\text{Hz}$$

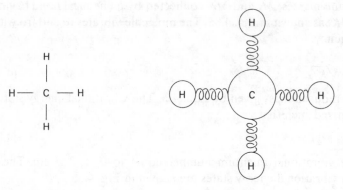

Figure 4.9   The mechanical analogue of a methane molecule.

In moving to a higher vibrational energy state, a molecule absorbs infrared (heat) radiation having a wavelength between 10^{-3} and 10^{-6} m. The source of infrared radiation is often an electrically heated bar of silicon carbide, emitting a continuous spectrum of radiation which has to be dispersed into its component wavelengths. Unfortunately glass does not transmit infrared radiation. A suitable monochromator is a prism of a metal salt such as NaCl or KBr which does not absorb in the infrared region. The dispersed radiation is then detected by a temperature-sensitive device, or thermometer. Suitable thermometers are the thermocouple and the thermistor which provide an electrical signal related to the heating effect of the radiation (see § 6.2).

Infrared absorption spectroscopy has found wide application in the analysis of organic molecules where functional groups such as $C=0$, $C-OH$ and $-CH_3$ tend to act as single masses and give characteristic absorption spectra. It is used in the analysis of paint solvents and polymeric materials and in the determination of some atmospheric pollutants.

4.7   Electron transitions in molecules

Radiation in the visible and ultraviolet regions of the spectrum may be absorbed when a molecular electron acquires energy to move to a higher available energy state, the wavelengths absorbed indicating the molecular electron energy level structure. Spectroscopically, visible and ultraviolet light act in the same way. However, since glass absorbs ultraviolet radiation, there are differences in the optics and the detectors used. Fortunately, quartz may be made into prisms which will transmit and disperse ultraviolet radiation. Similarly a quartz or fused silica sample holder is used. A suitable sample solvent is a paraffinic compound which does not absorb at these wavelengths. Sources of radiation are often tungsten filament or xenon discharge lamps which emit continuously over this part of the spectrum. The detector is a transducer which converts the radiant energy into an electrical signal. Typical detectors are the photocell and the photomultiplier tube which produce electric currents proportional to the intensity of the incident light (see § 11.6).

If the sample is in gaseous form, the electronic transition will in general be accompanied by simultaneous rotational and vibrational transitions during the absorption process. Absorption will occur at various closely spaced frequencies, depending on the accompanying rotational and vibrational energy changes. In an instrument with poor wavelength resolution, a broad absorption band results (Fig. 4.10a). With high resolution, the fine structure of the band may be analysed to provide information on the shape, flexibility and electronic structure of the molecule. Such an absorption band is shown in Fig. 4.10b for the toluene molecule, the band comprising lines due to different vibrational transitions, each vibrational line being subdivided into lines due to rotational transitions.

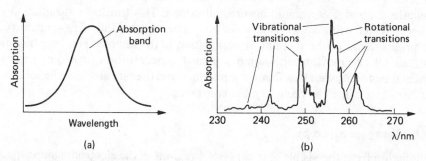

Figure 4.10 (a) A low resolution absorption band. (b) High resolution absorption band for toluene. (Adapted with permission from Robinson J. W., *Undergraduate Instrumental Analysis*, Marcel Dekker, 1970.)

Because the electron energy levels in a molecule depend largely on its atomic structure, the technique of electronic absorption spectroscopy has been applied to the analysis of polyatomic molecules such as vitamins, hormones and pesticides. One of the techniques of electron spectroscopy most useful in biology is concerned not with the absorption of radiation, but with its emission from excited atoms. If a valence electron in an atom is excited to a higher energy state it will return (within a time period of about 10^{-8} s) to the lowest available state, with the emission of a photon. The energy of the photon is equal to the energy difference between the states and is characteristic of the atom. Flame photometry is a technique used to measure the wavelength of the photon emitted when an atom is excited in a flame. The wavelength of the emission specifies the element present and its intensity indicates the abundance of this element. The method is particularly useful for determining concentrations of the alkali metals (Li, Na, K) and the alkali earth metals (Ca, Mg, Sr, Ba). The arrangement is shown in Fig. 4.11.

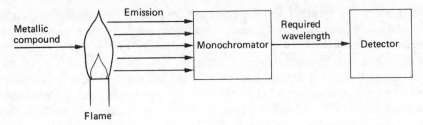

Figure 4.11 Flame photometry.

The metallic compound is dissolved in a suitable solvent and sprayed into a flame, where it vaporizes and is split into component atoms which are then thermally excited. A flame is a chemical reaction between two gases. A mixture of oxygen and hydrogen produces a flame temperature of 2800 °C. This is hot enough to excite the valence electrons of atoms such as sodium and potassium. The radiation which is subsequently emitted passes through the monochromator. For elements such as sodium which emits predominantly at a wavelength of 589 nm in the yellow region of the visible spectrum (more precisely, at two wavelengths; 589·0 nm and 589·6 nm), it is possible to place a filter between sample and detector which passes a narrow band of wavelengths around 589 nm and absorbs all others. This transmitted radiation is then detected using a photomultiplier tube. The emission intensity so measured is proportional to the metallic concentration in the sample. This may be estimated by comparison with some standard concentration providing experimental conditions such as flame temperature and the rate at which the solution is introduced into the flame remain constant.

4.8 X-ray spectroscopy

Radiation from the visible and ultraviolet regions of the electromagnetic spectrum has sufficient energy to excite atomic electrons to higher energy levels. That from the X-ray region, having greater energy, can remove electrons completely from the atom, leaving a positive ion. This mechanism is the basis of two spectroscopic techniques: X-ray absorption and X-ray fluorescence.

The source of radiation is the X-ray tube shown in Fig. 4.12 which comprises a cathode, in the form of a filament, and an anode, both in an evacuated container. When an electric current is passed through the cathode filament, it becomes hot and emits electrons. If the anode is at a potential V volts above the cathode, these electrons are accelerated towards the anode and reach it with appreciable kinetic energy. Since a charge of q coulombs accelerated through a potential V volts acquires energy qV joules, the kinetic energy of each electron at the anode is eV joules, where e is the electronic charge.

X-rays are produced when the energetic electrons strike the target of the anode. This is usually a button of material with high atomic number such as copper, molybdenum or tungsten. The X-ray emission spectrum of a molybdenum target is shown in Fig. 4.13a, where the intensity is proportional to the number of photons emitted at a particular wavelength. This typical spectrum has two components, the continuous spectrum extending over a wide

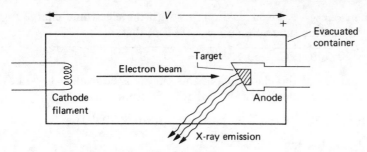

Figure 4.12 Schematic diagram of an X-ray tube.

range of wavelengths and the line spectrum where X-rays are emitted only at certain discrete wavelengths.

The continuous part of the spectrum arises from electron collisions with the target atoms, as shown in Fig. 4.13b. When an electron loses energy by this process, all or part of the energy lost may appear as a photon. Each photon emitted from the target arises from the slowing down of a single electron. The maximum energy of the continuous emission spectrum results from an electron losing all its energy in a single collision, this energy being emitted as a single photon. In this case

$$E_{max} = hf_{max} = eV \qquad (4.9)$$

Substituting $f = c/\lambda$ gives $hc/\lambda_{min} = eV$ or

$$\lambda_{min} = hc/eV \qquad (4.10)$$

where λ_{min} is the minimum wavelength emitted in the spectrum. This depends only on the tube voltage, whereas the intensity of emission is determined by the tube current, i.e. the number of electrons striking the target each second.

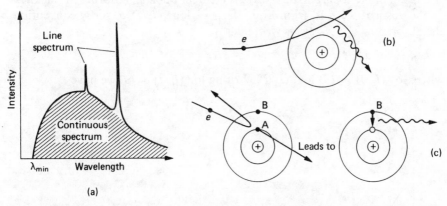

Figure 4.13 (a) The X-ray emission spectrum from a molybdenum target. (b) Origin of the continuous spectrum. (c) Origin of the line spectrum.

Example 4.6

Calculate the maximum energy and the corresponding minimum wavelength of X-rays from a tube operating at 50 kV.

The maximum energy is given by Equ. 4.9 with the electronic charge $e = 1 \cdot 60 \times 10^{-19}$C and the tube voltage $V = 50$ kV so that

$$E_{max} = 1 \cdot 60 \times 10^{-19}\text{C} \times 50 \times 10^3\text{V} = 8 \cdot 0 \times 10^{-15}\text{J}$$

From the definition of the electronvolt, this energy is equivalent to 50 keV. The corresponding wavelength of the radiation is, from Equ. 4.10

$$\lambda_{min} = 6 \cdot 63 \times 10^{-34}\text{J s} \times 3 \cdot 00 \times 10^8\text{m s}^{-1}/(8 \cdot 0 \times 10^{-15}\text{J})$$

$$\lambda_{min} = 2 \cdot 5 \times 10^{-11} \text{ m}$$

Relatively few electrons lose all their energy in a single collision, the main process being multiple collision loss giving photons of longer wavelengths. Similarly only about $0 \cdot 2\%$ of this energy loss is converted to X-rays, the remainder being dissipated as heat energy in the target. This is removed by circulating a coolant through the anode.

The sharp lines superimposed on the continuous spectrum arise from collisions of incident electrons with the electrons in the target atoms as shown in Fig. 4.13c. The incident electron removes an atomic electron A from an inner shell, leaving the target atom in an excited state. The atom de-excites when an electron B falls from an outer shell to fill the vacancy. In doing so it emits an X-ray of energy equal to the energy difference between the two states. This is characteristic of the target atoms so that the nature of the line spectrum depends on the target element used.

The detection of X-rays is based on their ability to ionize neutral atoms, so forming electron–ion pairs. If these are separated in an electric field, a charge may be collected proportional to the intensity of the radiation. Radiation detectors measuring the ionization produced by a beam of X-rays are described more fully in Chapter 11. A second type of detector is the photographic film, X-rays having a similar effect to light on a photographic emulsion. If a beam of monochromatic X-rays is incident upon an absorber of thickness L, the absorbance A is, from Equ. 4.2, $A = \log_{10}(I_0/I) \propto L$. By convention, the absorbance for X-rays is written as

$$A = \log_e(I_0/I)$$

where $\log_e(I_0/I) = 2 \cdot 3 \log_{10}(I_0/I)$. Writing $\log_e(I_0/I) = \mu L$ we have

$$I = I_0 e^{-\mu L}$$

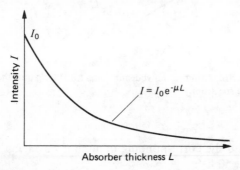

Figure 4.14   The variation of X-ray intensity through an absorber.

where μ is the linear absorption coefficient. This equation describes an exponential decrease in intensity with increasing thickness L as shown in Fig. 4.14. The linear absorption coefficient of a material increases with increasing atomic number and density, and varies with the energy of the incident X-rays. In general μ increases with decreasing energy.

Example 4.7

The linear absorption coefficients for X-rays of a particular energy in water and calcium are 400 m^{-1} and $1 \cdot 57 \times 10^4 \text{ m}^{-1}$ respectively. Calculate the percentage absorption of the X-rays in $1 \cdot 0$ mm of water and the thickness of calcium necessary to produce the same reduction in intensity.

Since $\log_e(I_0/I) = \mu L$, where $\mu = 400 \text{ m}^{-1}$ and $L = 1 \cdot 0 \times 10^{-3}$m for the water absorber, $\log_e(I_0/I) = 0 \cdot 40$ or $I_0/I = 1 \cdot 50$. Therefore the absorption is 33%. For calcium, $\log_e(I_0/I) = 0 \cdot 40 = \mu L$, and since $\mu = 1 \cdot 57 \times 10^4 \text{ m}^{-1}$

$$L = 0 \cdot 40/(1 \cdot 57 \times 10^4 \text{ m}^{-1})$$

or

$$L = 2 \cdot 5 \times 10^{-5} \text{m} = 25 \ \mu\text{m}$$

The increase in absorption with atomic number is the basis of the medical X-radiograph. Here, the 'sample' is part of the body and the detector is a photographic plate. Bone, which contains a large proportion of calcium, absorbs X-rays to a greater extent than soft tissue (largely carbon, hydrogen, oxygen and nitrogen) and its position is distinct. Conversely the region of the lungs is evident because of the poor absorption of low density air. Sometimes it may be necessary to radiograph parts of the body without such a fortunate elemental distribution, for example the gastro-intestinal tract. This problem may be overcome by allowing the subject to drink a compound of barium which quickly distributes itself throughout the tract and is a very efficient absorber of X-rays. Using the continuous-energy X-ray spectrum it is possible to determine the calcium content of bone. This type of measurement has recently been extended to soft tissue. At low energies, the linear absorption coefficients for lean and fat tissue are slightly different, and by comparing the absorbances of tissue to X-rays of two different energies the lean-to-fat ratio at various anatomical sites can be measured.

The second major technique in X-ray spectroscopy is X-ray fluorescence. When a sample is placed in a beam of X-rays, some radiation is absorbed in causing atomic excitations. Fluorescent X-rays are then emitted as the atom de-excites which have an energy characteristic of the element involved. Measurement of the intensity and energy of the emission determines the type and amount of a particular element in the sample. This technique has been used to measure lead levels in blood and strontium in tissue, and recent applications include the elemental analysis of wood and food. The advantages of this technique are that little sample preparation is required and the sample is not altered chemically. In principle X-ray fluorescence may be used to measure all but the lightest elements although in practice it would not be used for determinations of elements such as sodium, for which flame photometry is more convenient.

Problems

4.1   A molecule absorbs radiation of frequency $2 \cdot 0 \times 10^{13}$ Hz. What is the energy difference, in eV, between initial and final states and what type of molecular transition is involved?

4.2   In the emission spectrum of the sun, the relative intensities of photons emitted at wavelengths 650 nm and 500 nm is $1 \cdot 13:1$. Find the ratio of the total energies radiated at the two wavelengths.

4.3   If two samples having absorbances A_1 and A_2 are placed together show that the absorbance of the combination is $A = A_1 + A_2$. A sample is placed in a beam of radiation and the absorbance is measured as $0 \cdot 125$. A similar sample containing the same solution is then placed in the beam immediately behind the first. If the original intensity of the radiation was 500 units, calculate the intensity of the radiation leaving the combination.

4.4   At a certain wavelength, the absorbance of a solution of length 10 mm is found to be equal to that of a standard of the same solution of length 40 mm. If the standard concentration is 100 ppm, find the concentration of the unknown. If the two solutions were of different molecular composition, would it be possible to determine this concentration?

4.5   Sodium chloride has a bond length of $0 \cdot 236$ nm. Calculate the reduced mass and the moment of inertia for this molecule and find the frequency of the radiation absorbed in a rotational transition from state $J = 0$ to $J = 1$.

4.6   Using results derived in Example 4.3, calculate the energies of the first four rotational levels in the NO molecule.

4.7   The carbon monoxide molecule absorbs radiation of frequency $1 \cdot 93 \times 10^{14}$ Hz in a vibrational transition from state $v = 0$ to $v = 3$. Calculate the force constant of the $C - O$ bond. What is the energy difference, in eV, between successive vibrational levels?

4.8   The amplitude of molecular vibrations may be determined since, at the turning point of the vibration, the potential energy $Ky_0^2/2$ is equal to the total energy of the state. Using the value of vibration frequency determined in Example 4.5, find the vibration energy of the $v = 0$ state of the $C - H$ bond and the vibration amplitude in this state.

4.9   The following data were obtained to prepare a calibration curve to determine potassium concentrations by flame photometry. Use the calibration curve to find the concentrations of samples giving emission intensities of $8 \cdot 3$, $17 \cdot 8$, $26 \cdot 0$, $30 \cdot 1$ and $38 \cdot 6$ units.

K concentration/ppm	$0 \cdot 2$	$0 \cdot 4$	$0 \cdot 6$	$0 \cdot 8$	$1 \cdot 0$	$1 \cdot 2$
Emission intensity/units	$6 \cdot 6$	$13 \cdot 6$	$20 \cdot 2$	$26 \cdot 9$	$33 \cdot 8$	$40 \cdot 5$

4.10   Electrons emitted by the filament in an X-ray tube have potential energy. Describe the subsequent energy transformations in the process of X-ray production.

4.11   The concentration of an element in a sample may be determined by X-ray fluorescence, where atoms are excited by X-rays and emit characteristic X-rays which may be deteced. Copper atoms emit such fluorescent X-rays of wavelength 0·154 nm. What is the minimum tube voltage required to excite this fluorescence? A 0·10 mm aluminium filter reduces the intensity of the fluorescent radiation by 74%. What is the linear absorption coefficient of aluminium for the radiation?

5

Electricity

Electricity is important for two reasons: (a) complex organisms such as animals are controlled by electrical signals derived from sensors which respond to changes in the environment; and (b) electrical and electronic circuits are widely used in biological measurement. In this chapter we describe the nature of electrical signals and the behaviour of simple passive circuits. The way in which more complex electronic instruments are used for measurement is discussed in Chapter 6.

5.1 Current and voltage

We have seen in Chapter 3 that all matter contains electrically charged particles. Electrical phenomena are due to the interaction and movement of these particles. If a free charged particle with charge q is situated in an electric field of intensity E it will experience a force $F = Eq$ so that it will be accelerated (Equ. 1.18). If the particle carries positive charge it moves in the direction of the field, from $+$ to $-$, as shown in Fig. 5.1a, and if it carries negative charge it moves in the opposite direction, from $-$ to $+$. The continuous movement of charges under the influence of an electric field is called electrical conduction.

There are two broad categories of materials: electrical conductors, in which there are a large number of free charge carriers; and electrical insulators (e.g.

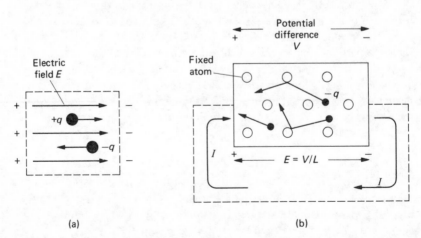

Figure 5.1 (a) The motion of charged particles in an electric field. (b) The collision of moving charged particles, $-q$, with lattice atoms in an electrical conductor.

plastics) in which the charge carriers are fixed and cannot move. The two types of electrical conductor of particular interest are materials containing free electrons (the metals) and materials containing relatively free ions (electrolyte solutions). When metal atoms combine to form solid material the valence electrons become detached from the atoms (§ 3.4). These negatively charged electrons are free to move within the material under the influence of an electric field. This type of conduction is called electronic conduction. When acids, bases and salts are put into solution the molecules dissociate to form positively and negatively charged ions (§ 3.4). For example, when NaCl is dissolved in water the electrically neutral NaCl molecule dissociates to form a positively charged sodium ion, Na^+, and a negatively charged chloride ion, Cl^-. Both of these ionic species are free to move through the solution and conduct electricity. The Na^+ ions move in one direction (from + to −) and the Cl^- ions move in the opposite direction. This is called ionic conduction.

The movement of charged particles in a particular direction constitutes a current. In an electrical circuit the current is due to the flow of electrons (electronic current) whereas in living organisms it is due to the flow of ions (ionic current). For a current to flow the conducting system must be in the form of a closed loop. The movement of negative charges in a material is shown in Fig. 5·1b. The closed loop is indicated by the dashed line. This is called an electric circuit.

The flow of negative charges in a particular direction, say from right to left, is equivalent to the flow of positive charges in the opposite direction. The direction of current flow, which is indicated by an arrow (with symbol I), shows the direction of movement of positive charges, or its equivalent.

The current I is defined as the amount of charge that passes through a circuit in one second:

$$I = \Delta Q / \Delta t \tag{5.1}$$

where ΔQ (coulombs) is the amount of charge that flows in time Δt (seconds). The unit of current is the ampere ($C\ s^{-1}$) which is often abbreviated amp (A).

The movement of charge carriers through a material is impeded by collisions with the atoms or molecules of the medium (Fig. 5.1b). In order to overcome this resistance to charge movement and obtain a steady flow of current it is necessary to establish a potential difference across the conducting medium. The potential difference V, which has units of volt (V), creates an electric field of intensity E ($N\ C^{-1}$) which causes the charge carriers to move. E is related to V by $E = V/L$ where L is the length of the conductor. The units $V\ m^{-1}$ are equivalent to $N\ C^{-1}$. With many electrical conductors the potential difference is directly proportional to the current:

$$V = IR \tag{5.2}$$

The constant of proportionality R is called the electrical resistance and has units of ohm ($\Omega = V\ A^{-1}$). Equ. 5.2 is known as Ohm's law. The resistance of a conductor is proportional to its length L and inversely proportional to its cross-sectional area A

$$R = \rho L/A \tag{5.3}$$

where ρ, the resistivity of the material, has units of Ω m. The reciprocal of ρ is called the conductivity. For metals the resistivity increases with increasing temperature because the number of collisions that an electron makes with the lattice atoms increases.

Example 5.1

The tip of a micropipette electrode used for the measurement of cell potentials consists of a glass capillary of length 2·0 mm and inside diameter 2·0 μm. The inside of the capillary is filled with a saturated NaCl solution of resistivity $4·0 \times 10^{-2}\Omega$ m. What is the resistance of the column of electrolyte?

From Equ. 5.3, $R = \rho L/A = \rho L/\pi r^2$ so that

$$R = \frac{4·0 \times 10^{-2}\Omega \text{ m} \times 2·0 \times 10^{-3}\text{m}}{3·14 \times (1·0 \times 10^{-6} \text{ m})^2}$$

$$R = 2.5 \times 10^7\Omega = 25M\Omega$$

The electrical potential at a point in a circuit is always measured relative to some reference level (in the same way as mechanical potential energy is measured with respect to a reference, § 1.5). This is usually taken as earth, or ground, which is defined as zero potential (0 V). The symbols used to represent ground are shown in Fig. 5.2a. The potential difference, V, across a conductor is the difference in the potentials V_a and V_b at the two ends, $V = V_a - V_b$. Current, I, always flows from a higher potential to a lower potential, i.e. from $+$ to $-$. The unit of potential difference is the volt and the terms potential difference and voltage are used synonymously.

Whenever electrical charges of opposite sign become separated, as in Fig. 5.2b, an electric field is created and a potential difference, V, is established. This separation of charge can occur for several reasons, e.g. temperature gradients (as in the thermocouple, § 6.2) or concentration gradients (as in a nerve

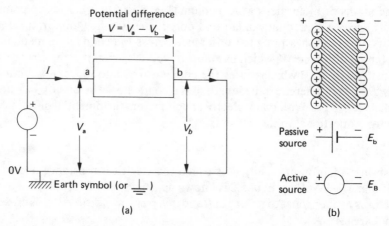

Figure 5.2 (a) The potentials at different points in a circuit. (b) The potential difference due to separation of charged particles into an electrical double layer.

fibre, § 5.4). If the charges exist on the surface of a material, the material is said to be polarized. There are two types of source of potential difference: passive and active. The symbols used to represent these sources are shown in Fig. 5.2b. A passive source, such as a battery, produces a steady potential difference across its electrodes. Passive sources are used to provide power in electrical circuits. An active source, such as a muscle, produces a potential difference which is related to some activity. When a current or a voltage is derived from an active source it is referred to as the signal. The origin of the signal is often indicated by a subscript, e.g. E_B, biological origin, E_T, thermal origin. All sources have internal resistance. This is ignored in the following discussion of electrical circuits.

The important quantities that describe a signal are its size (volt or amp) and its time dependence. There are two kinds of current, direct current (DC) in which the charges always flow in the same direction and alternating current (AC) in which the charge flows backwards and forwards. Various combinations of DC and AC currents exist some of which are shown in Fig. 5.3. The alternating current I' shown in Fig. 5.3b can be described by the equation $I' = I_0 \sin(2\pi ft)$ where I_0 is the amplitude of the signal current and f the frequency. If this current flows through a conductor of resistance R, the voltage across the conductor is, from Ohm's law, $V' = I_0 R \sin(2\pi ft) = V_0 \sin(2\pi ft)$.

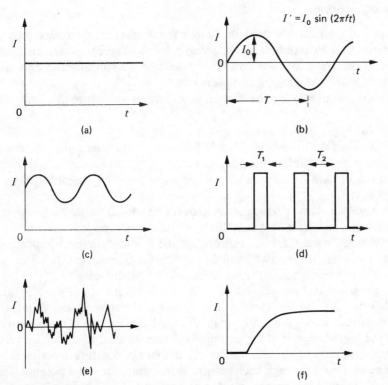

Figure 5.3 Different types of signal current: (a) direct current (DC); (b) alternating current (AC); (c) pulsating current, a combination of (a) and (b); (d) pulses with pulse duration T_1, and mark-space ratio T_1/T_2; (e) random signal; (f) a single event.

The voltage signal has the same waveform as the current and has amplitude $V_0 = I_0 R$.

When a current flows through a conductor work is done in moving the charges from one location to another so that in order to get current flow an energy source, i.e. a battery or an active signal source, must be provided. The rate at which energy is used up is given by

$$P = VI = I^2 R = V^2/R \qquad (5.4)$$

where P(watts) is the power dissipated in the conductor. The electrical energy is transformed into heat energy and the conductor becomes hot (see §1.5).

When an alternating current I' flows through a conductor the power dissipated is the time-average value of the product $I'V'$. In this case

$$P = (I_0/\sqrt{2})(V_0/\sqrt{2}) = I_{rms} V_{rms} = I^2_{rms} R$$

where I_0 and V_0 are the amplitudes of the current and voltage. $I_0/\sqrt{2}$ is called the root-mean-square current, I_{rms}, and $V_0/\sqrt{2}$ is V_{rms}. The voltage obtained from AC power supply systems is usually described in terms of the RMS voltage. A 220 V AC mains voltage means 220 V_{rms} so that the amplitude of the AC mains voltage is 310 V.

5.2   Circuit elements

Two types of element are used in electrical and electronic circuits: passive and active. Passive elements impede current flow whereas active elements generate a change in current flow. In this section we consider three passive elements: resistors, capacitors and inductors.

Ohm's law (Equ. 5.2), can be written in a more general form as

$$V = IZ \qquad (5.5)$$

where Z is called the impedance. The impedance of an element for AC current is not necessarily the same as that for DC current. A typical appearance, symbol, unit and the DC and AC impedances of the three passive elements are shown in Fig. 5.4.

A resistor is an element which offers a specific electrical resistance to current flow. Resistors are usually made from a carbon composite or from high resistivity metal wire. Carbon resistors are used in low current applications (e.g. electronic circuits) and wire wound resistors for high currents (e.g. heating elements). The impedance is the same for AC and DC currents.

Most electrical circuits can be analysed using Ohm's law (Equ. 5.5), and remembering that: (a) the current in a circuit is the same at any point; (b) the total current into a junction equals the total current out of the junction; and (c) the algebraic sum of the potential differences around a circuit is zero. The simplest electrical circuit consists of a number of resistors in series, (Fig. 5.5a). The current I through each resistor is the same. The total potential difference, E_b, across the arrangement is the sum of the potential differences across each element, $V_1 = IR_1$, $V_2 = IR_2$, $V_3 = IR_3$ so that $E_b = IR_1 + IR_2 + IR_3$. The effective resistance of the three resistors in series is the total potential

Element	Symbol	Unit	Impedance $|Z|$	
			DC	AC
(a) 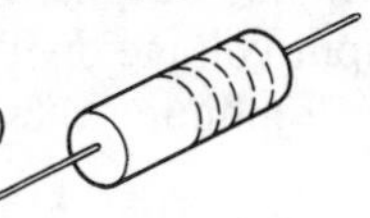	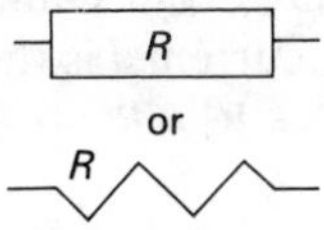 R or R	ohm (Ω)	R	R
(b) 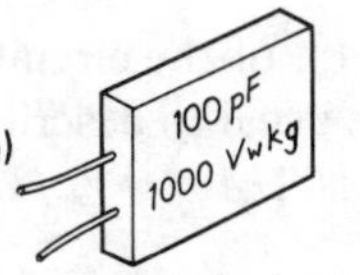	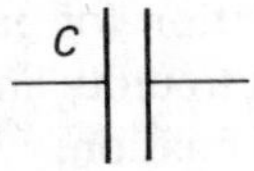 C	farad (F)	∞	$\dfrac{1}{2\pi f C}$
(c) 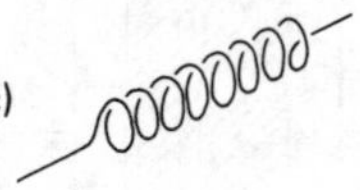	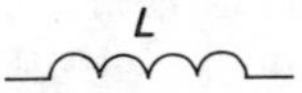 L	henry (H)	0	$2\pi f L$

Figure 5.4 The three passive elements: (a) resistor, (b) capacitor and (c) inductor.

difference E_b divided by the current I:

$$R = E_b/I = R_1 + R_2 + R_3 \tag{5.6}$$

When resistors are arranged in parallel (Fig. 5.5b) there are three separate circuits, the potential difference across each circuit being E_b. The total current I flowing through this arrangement is the sum of the currents, $I_1 = E_b/R_1$, $I_2 = E_b/R_2$ and $I_3 = E_b/R_3$, in each circuit so that $I = E_b/R_1 + E_b/R_2 + E_b/R_3$. The effective resistance, $R = E_b/I$, of the parallel combination is

$$\frac{1}{R} = \frac{1}{R_1} + \frac{1}{R_2} + \frac{1}{R_3} \tag{5.7}$$

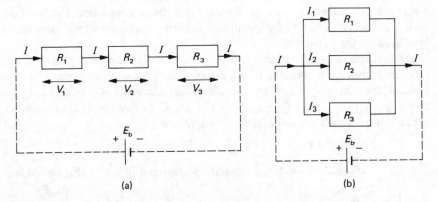

Figure 5.5 (a) Resistors in series. (b) Resistors in parallel.

A commonly encountered circuit is shown in Fig. 5.6a. This is called a potential divider network. A voltage V_{in} is applied to the input of the circuit and an output voltage V_{out} is taken across the resistor R_2. The input voltage causes a current I to flow through the circuit. From Ohm's law the output voltage $V_{out} = IR_2$. Again from Ohm's law the current I is the input voltage divided by the total resistance of the circuit, $I = V_{in}/(R_1 + R_2)$. Combining these two equations gives the output voltage as

$$V_{out} = V_{in} R_2/(R_1 + R_2) \qquad (5.8)$$

The ratio V_{out}/V_{in} is called the amplification (or gain), A_V, of the circuit. For passive circuits A_V is less than or equal to unity and it is usual to describe the behaviour of these circuits by the attenuation, which is $1/A_V = V_{in}/V_{out}$.

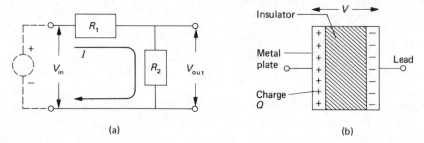

(a)   (b)

Figure 5.6   (a) A potential divider network. (b) A charged capacitor.

The simplest form of capacitor consists of two metal plates separated by an electrical insulator (Fig. 5.6b). When charges of opposite sign are established on the plates, say by momentarily connecting them to the terminals of a battery, the mutual attraction between unlike charges (§ 1.4) holds the charge on the plates. Thus a capacitor has the capacity for storing charge. The amount of charge Q (coulomb) stored on one plate is proportional to the potential difference V (volt) between the plates:

$$Q = CV \qquad (5.9)$$

where the constant of proportionality C is called the capacitance. The unit of capacitance is the farad F = C V^{-1}). The potential difference at which electrical breakdown of the insulating layer occurs is called the working voltage. The values of the capacitance and the working voltage are usually indicated on the capacitor.

When capacitors are connected in parallel (Fig. 5.7a) the potential difference across each element is the same. The total charge stored Q is the sum of the charges stored by the individual elements, $Q_1 = C_1E_b$, $Q_2 = C_2E_b$, $Q_3 = C_3E_b$. Thus $Q = C_1E_b + C_2E_b + C_3E_b$ and the total capacitance for the three capacitors in parallel, $C = Q/E_b$, is

$$C = C_1 + C_2 + C_3$$

The total capacitance for three capacitors arranged in series (Fig. 5.7b) is

$$\frac{1}{C} = \frac{1}{C_1} + \frac{1}{C_2} + \frac{1}{C_3}$$

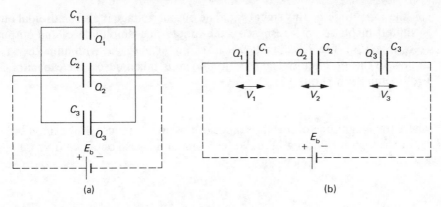

Figure 5.7   (a) Capacitors in parallel. (b) Capacitors in series.

Since the metal plates of a capacitor are separated by an insulator the impedance to DC current flow is infinite. For sinusoidal AC currents the magnitude of the impedance is $|Z_C| = 1/2\pi f C$ so that a capacitor offers little impedance to the flow of high frequency current.

An inductor consists of a number of turns of insulated wire wound in the form of a coil, as shown in Fig. 5.4c. When a current I flows through this coil a magnetic field is produced which is proportional to I. This field passes through the coil. If the current changes, the magnetic field changes and a potential difference is induced across the ends of the coil. The induced voltage V is proportional to the time rate of change of the current, $V = -L(\Delta I/\Delta t)$, where the constant of proportionality, L (henry), is called the self-inductance of the coil. The minus sign indicates that the induced potential difference opposes the change in current. The inductor is constructed of high conductivity wire so that its impedance to DC current flow is zero. However, because of the magnetic interaction when the current changes, it offers impedance to AC current. For a sinusoidal current, $I' = I_0 \sin(2\pi f t)$, the magnitude of the impedance is $|Z_L| = 2\pi f L$.

5.3   Combinations of circuit elements

Various combinations of circuit elements occur in measurement systems which influence the signals produced by the system. These elements may be included by design, to produce a controlled modification of the signal, or they may occur unintentionally and may produce an unwanted modification of the signal. In either case it is necessary to know how combinations of passive elements influence current and voltage signals. In biological measurement the signals of most interest are sinusoidal (AC) voltages and pulses.

Two commonly encountered circuits are the low-pass filter, which allows the passage of low-frequency signals but attenuates high-frequency signals and the high-pass filter which is the reverse of this. A low-pass filter consists of an RC combination as shown in Fig. 5.8. The AC input voltage is of the form $V = V_0 \cos(2\pi f t)$ where V_0 is the amplitude and f the frequency. We will call the amplitude of this signal V_{in}. The output voltage is also sinusoidal but is shifted in phase with respect to the input. The amplitude of the output voltage is V_{out}. The input voltage causes an AC current, with amplitude I_0, to flow in the RC circuit. The output voltage is taken across the capacitor C. From Ohm's law (Equ. 5.5),

$$V_{out} = I_0|Z_C|$$

where the magnitude of the impedance of C is given in Fig. 5.4. The amplitude, I_0, of the current flowing in the RC combination is also obtained from Ohm's law.

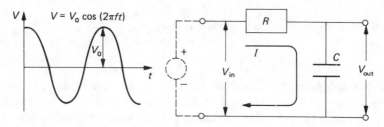

Figure 5.8   A low-pass filter with sinusoidal input.

When dealing with AC quantities the total impedance is not simply the sum of the individual impedance magnitudes; account must be taken of the phase relationship between the voltage and the current. If a potential difference $V = V_0 \cos(2\pi ft)$ exists across a resistor the current I flowing through the resistor, $I = (V_0/R) \cos(2\pi ft)$, is in phase with the voltage V and has amplitude $I_0 = (V_0/R)$. When the same voltage exists across a capacitor the charge Q on the capacitor is $Q = CV_0 \cos(2\pi ft)$. Since $I = \Delta Q/\Delta t$ the current through the capacitor can be obtained by taking the slope of the voltage – time graph, $I = CV_0\Delta[\cos(2\pi ft)]/\Delta t$. This was done in Section 2.1 to obtain velocity from position. It can be shown that $I = -2\pi fCV_0 \sin(2\pi ft)$. The phase relationship between the current and the voltage is the same as that between the waveforms of Fig. 2.2c and b. Since $-\sin\theta = \cos(\theta + \pi/2)$ the equation for I can be written as $I = 2\pi fCV_0 \cos(2\pi ft + \pi/2)$. The current through the capacitor leads the voltage across the capacitor by 90 degrees (or the voltage lags the current by 90 degrees). The amplitude of the current $I_0 = 2\pi fCV_0$.

In the circuit of Fig. 5.8 the AC current I, of amplitude I_0, flows through both elements. The voltage V_R across R is in phase with I and the voltage V_C across C lags I by 90 degrees. These voltages can be represented by lines in a two-dimensional coordinate system as shown in Fig. 5.9. The lengths of the lines are proportional to the magnitudes of the voltages, $|V_R| = I_0 R$ and $|V_C| = I_0/2\pi fC$, and the directions of the lines are given by the phase differences between the voltages and the current, zero and -90 degrees respectively. The magnitude $|V'|$ of the voltage across the combination is the resultant of the two components, $|V'| = \sqrt{[|V_R|^2 + |V_C^2|]} = I_0\sqrt{[R^2 + 1/(2\pi fC)^2]}$. Defining the magnitude of the impedance of the series RC combination as $|Z| = |V'|/I_0$ gives

$$|Z| = \sqrt{[R^2 + 1/(2\pi fC)^2]}$$

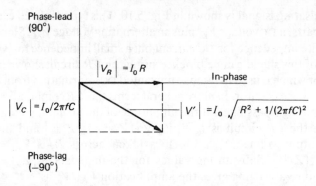

Figure 5.9   The addition of the voltages across the elements of the RC combination.

The amplitude I_0 of the current in the low-pass filter is related to the amplitude V_{in} of the input voltage by

$$I_0 = \frac{V_{in}}{|Z|} = \frac{V_{in}}{\sqrt{[R^2 + 1/(2\pi fC)^2]}}$$

Since $V_{out} = I_0|Z_C|$ and $|Z_C| = 1/2\pi fC$ we have

$$V_{out} = \frac{V_{in}(1/2\pi fC)}{\sqrt{[R^2 + 1/(2\pi fC)^2]}} = \frac{V_{in}}{\sqrt{[1 + (2\pi fRC)^2]}}$$

and the voltage ratio

$$\frac{V_{out}}{V_{in}} = \frac{1}{\sqrt{[1 + (2\pi fRC)^2]}} \qquad (5.10)$$

For low-frequency signals, as f tends to zero, V_{out}/V_{in} is equal to unity so that the attenuation is unity. For high-frequency signals, as f tends to infinity, V_{out}/V_{in} is zero so that the attenuation is infinite. f tends to infinity means that the product $f \times RC$ is much larger than unity.

Example 5.2

An electrical signal obtained from a measurement system consists of a low-frequency biological signal plus signals at high frequency which are of no interest. These high-frequency signals must be attenuated using an RC (low-pass) filter. If $R = 1 \cdot 0$ MΩ what is the value of C which will give an attenuation of 10 at 20 Hz?

From Equ. 5.10, the attenuation $V_{in}/V_{out} = \sqrt{[1 + (2\pi fRC)^2]}$. Squaring both sides and re-arranging gives $(RC)^2 = [(V_{in}/V_{out})^2 - 1]/(2\pi f)^2$. Substituting values

$$(RC)^2 = \frac{100 - 1}{(2 \times 3 \cdot 14 \times 20 \text{ s}^{-1})^2} = 6 \cdot 3 \times 10^{-2} \text{ s}^2$$

$$RC = 2 \cdot 5 \times 10^{-1} \text{ s}$$

Substituting for R gives $C = 2 \cdot 5 \times 10^{-1}\text{s}/(1 \cdot 0 \times 10^6 \Omega) = 0 \cdot 25 \ \mu\text{F}$.

A more complex, pulsating, signal is shown in Fig. 5.10. This type of signal can be divided into two parts: a DC voltage V_{DC} plus an alternating voltage V_{AC}. Since a capacitor has infinite impedence for DC current but a small impedence for AC current the DC part of this signal can be blocked using the CR circuit shown in Fig. 5.10. A capacitor which is used to remove unwanted DC information from a signal is called a blocking capacitor. In more general terms this CR combination can be considered as a high-pass filter. If the amplitude of the AC input voltage is V_{in} the current in the CR circuit is $I_0 = V_{in}/\sqrt{[|Z_R|^2 + |Z_C|^2]}$ and the amplitude of the output voltage V_{out}, which is taken across R, is $V_{out} = V_{in}|Z_R|/\sqrt{[|Z_R|^2 + |Z_C|^2]}$. Substituting values for the impedances $|Z_R| = R$, $|Z_C| = 1/2\pi fC$ and re-arranging gives the amplification V_{out}/V_{in} as

$$\frac{V_{out}}{V_{in}} = \frac{R}{\sqrt{[R^2 + 1/(2\pi fC)^2]}} = \frac{1}{\sqrt{[1 + 1/(2\pi fRC)^2]}} \tag{5.11}$$

When f is zero, $V_{out}/V_{in} = 0$, and when f is infinite, $V_{out}/V_{in} = 1$ i.e. $V_{out} = V_{in}$.

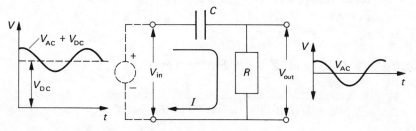

Figure 5.10   A high-pass filter with a pulsating input voltage.

The way in which the amplification of these two types of filter varies with frequency, which is called the frequency response, is shown in Fig. 5.11. In both cases the amplification $V_{out}/V_{in} = 1/\sqrt{2} = 0\cdot707$ when the frequency $f = 1/2\pi RC$. This frequency is called the cut-off frequency. For the low-pass filter it is the upper cut-off frequency, f_u, and for the high-pass filter it is the lower cut-off frequency, f_l.

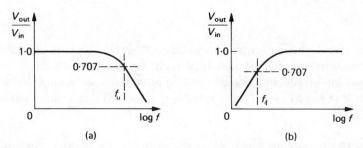

Figure 5.11   The frequency response of (a) a low-pass filter and (b) a high-pass filter.

Another important aspect of RC circuits is how they affect a voltage pulse. A positive voltage pulse of duration T_1 may be considered as a voltage step V', at $t = 0$, followed T_1 seconds later by a voltage step $-V'$, as shown in Fig. 5.12a. The circuits of Figs. 5.8 and 5.10 are basically the same, except that in Fig. 5.8 the output voltage is taken across the capacitor whereas in Fig. 5.10 it is taken across the resistor. The generalized circuit is shown in Fig. 5.12d. The voltage V_{in} at the input is equal to the sum of the voltage drops across R and C, i.e. $V_{in} = V_R + V_C = IR + Q/C$. At times less than zero $V_{in} = 0$ and $V_R = V_C = 0$. At the leading edge of the pulse, $t = 0$, the voltage V_{in} suddenly increases to V'. At this instant the charge on the capacitor is zero so that $V_C = 0$ (Fig. 5.12b). (An alternative way of considering this situation is to note that the voltage step V' is equivalent to a very high-frequency signal. Since the impedance $|Z_C|$ of the capacitor to high-frequency signals is zero the voltage drop across C is zero.) At time greater than zero, charges $+Q$ and $-Q$ accumulate on the two plates of the capacitor so that $V_C (= Q/C)$ increases. It can be shown that during the charging of the capacitor, with constant input voltage V', the voltage V_C is

$$V_C = V'(1 - e^{-t/\tau}) \tag{5.12}$$

where $\tau = RC$ is called the time constant of the circuit. The voltage increases with time and eventually reaches an equilibrium value V' when t/τ is much greater than unity, i.e. $t \gg \tau$. The voltage reaches a value of $V'(1 - e^{-1}) = 0.63\ V'$ after a time $t = \tau$. The smaller the value of τ the more rapidly the capacitor charges.

At the trailing edge of the pulse ($t = T_1$) the voltage suddenly drops to zero. The positive charge, $+Q$, flows from the positive plate through R and the external circuit to the negative plate and neutralizes the negative charge, $-Q$, so that the voltage V_C decreases. This is called the discharge of the capacitor and

$$V_C = V'e^{-t/\tau} \tag{5.13}$$

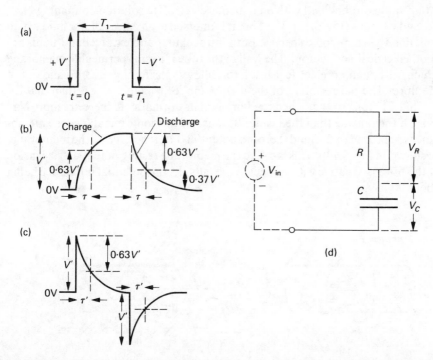

Figure 5.12 (a) The input pulse. (b) The voltage across C. (c) The voltage across R. (d) The generalized RC circuit.

The voltage drops to $0 \cdot 37 \, V'$ after a time $t = \tau$ and eventually falls to zero when $t \gg \tau$.

The voltage across the resistor, $V_R = V_{in} - V_C$, is shown in Fig. 5.12c. At $t = 0$, $V_C = 0$ so that $V_R = V_{in} = V'$. At times greater than zero $V_R = V' - V'(1 - e^{-t/\tau'}) = V'e^{-t/\tau'}$ so that V_R decreases exponentially with time and eventually falls to zero. (Although τ' is numerically equal to RC we have made a distinction between τ and τ' because the circuits shown in Fig. 5.8 and 5.10 affect a voltage pulse in different ways.) When the capacitor is completely charged $V_C = V'$ so that $V_R = 0$. At time $t = T_1$ the voltage V_{in} falls to zero so $V_R = 0 - V_C = -V'$, i.e. V_R rapidly decreases to $-V'$. It then returns to zero exponentially with time.

Figs. 5.12b and c represent the response of the low-pass filter (Fig. 5.8) and the high-pass filter (Fig. 5.10) respectively to a voltage pulse of size V' and duration T_1. To obtain an undistorted output voltage from a low-pass filter its time constant τ should be much smaller than T_1. For a high-pass filter τ' must be much larger than T_1 for the output to be undistorted. The pulse response of these circuits is intimately related to their frequency response and it can be shown that $\tau = 1/2\pi f_u$ and $\tau' = 1/2\pi f_l$.

5.4 The passive electrical properties of nerve fibres

Certain types of cell within living organisms, such as nerve and muscle fibres, are able to respond to and propagate electrical signals. A nerve fibre consists of a thin hollow tube filled with an aqueous electrolyte solution containing Na^+, K^+ and Cl^- ions (Fig. 5.13). The fibre is immersed in an extracellular fluid which contains the same ionic species. Both intracellular and extracellular fluids are good electrical conductors. The wall of the tube is a semi-permeable membrane which, although an electrical insulator, allows ions to migrate into and out of the fibre. The permeability of the membrane for Na^+ ions is less than that for K^+ and Cl^- ions so that when the fibre is in its normal state there are more Na^+ ions on the outside than the inside. Because of this ionic imbalance the membrane is polarized (§ 5.1) and there is a potential difference across the membrane, as shown in Fig. 5.13b. This is called the membrane resting potential. The inside of the fibre is negative with respect to the outside, the potential difference being about 90 mV.

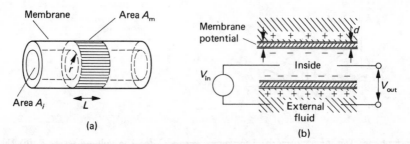

Figure 5.13   A nerve fibre.

The electrical properties of a fibre can be determined by applying a voltage signal, V_{in}, to one end of a short length of the fibre and measuring the voltage, V_{out}, at the other (Fig. 5.13b). The way in which this is done is described in Chapter 6. When a small positive voltage pulse, V_{in}, of duration 2 ms and size about 20 mV is applied to the fibre the output voltage V_{out} is smaller in size than V_{in} and is distorted. The attenuation increases with increasing length of fibre. These results indicate that for small exciting voltages the short length of fibre behaves like the passive electrical circuit shown in Fig. 5.14 which contains resistance and capacitance. An equivalent circuit has the same response as the real system. R_m and C_m represent the transverse (through the thickness) resistance and capacitance of the membrane and R_i the longitudinal resistance of the fibre interior.

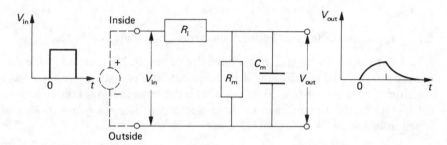

Figure 5.14 The input and output voltages for a short length of nerve fibre and the equivalent circuit.

Ignoring the capacitor, C_m, the equivalent circuit is the same as the potential divider network of Fig. 5.6a so that the attenuation produced by the short length of fibre is, from Equ. 5.8, $V_{in}/V_{out} = (R_m + R_i)/R_m$. Measurement shows that for a particular nerve fibre the resistivity ρ_m of the cell membrane is $4.0 \times 10^7 \Omega$ m and the resistivity ρ_i of the cell interior is 0.60Ω m. For a small piece of fibre 2.0 mm long, radius r approximately 40 μm and membrane thickness d about 50×10^{-10} m the resistance R_i of the interior is, from Equ. 5.3, $R_i = \rho_i L/A_i = \rho_i L/\pi r^2$ so that

$$R_i = \frac{0.60 \ \Omega \ \text{m} \times 2.0 \times 10^{-3} \ \text{m}}{3.14 \times 1.6 \times 10^{-9} \ \text{m}^2} = 2.4 \times 10^5 \ \Omega$$

The resistance R_m of the fibre membrane is $R_m = \rho_m d/A_m = \rho_m d/2\pi rL$

$$R_m = \frac{4.0 \times 10^7 \ \Omega \ \text{m} \times 50 \times 10^{-10} \ \text{m}}{2 \times 3.14 \times 40 \times 10^{-6} \ \text{m} \times 2.0 \times 10^{-3} \ \text{m}} = 4.0 \times 10^5 \ \Omega$$

The attenuation in this short length of fibre is

$$\frac{V_{in}}{V_{out}} = \frac{2.4 \times 10^5 \ \Omega + 4.0 \times 10^5 \ \Omega}{4.0 \times 10^5 \ \Omega} = 1.6$$

The effective capacitance of the cell membrane can be determined from the time constant τ of the output voltage (Problem 5.11).

This calculation shows that the signal is reduced to about 0.63 of its initial value when it propagates 2 mm along the fibre. A signal from the central nervous system would also be attenuated by the same amount and would rapidly disappear. The passive attenuation in a nerve fibre is overcome by an active mechanism. If a voltage pulse greater than about 30 mV is applied to the fibre the membrane becomes much more permeable to Na^+ ions and these ions rapidly migrate from the extracellular fluid into the interior of the fibre. This causes the potential on the inside of the membrane to rise and it eventually becomes positive with respect to the outside. This process is called depolarization. After a short time the flow of Na^+ ions stops, the membrane becomes impermeable to these ions and returns to its initial resting state. The voltage pulse produced by the depolarization and the repolarization of the fibre membrane is called the action potential.

Problems

5.1   Explain the difference between electronic and ionic conduction.

5.2   Four micropipette electrodes have tips of length 2·0 mm and inside diameters 1·0, 5·0, 10 and 20 μm. If the capillary is filled with a saturated NaCl solution of resistivity $4·0 \times 10^{-2}$ Ω m what is the resistance R_E of the column of electrolyte in the tips of each of these electrodes? Draw a graph of R_E as a function of $\log_{10} r$.

5.3   What are the resistances of pieces of silver wire, resistivity $1·6 \times 10^{-8}$ Ω m, which have the same dimensions as the electrolyte columns of Problem 5.2?

5.4   A 240 V_{rms} mains power supply outlet socket is fitted with a 5 amp fuse. Explain what happens when two 1 kW heating mantels wired in parallel are connected to this socket.

5.5   A 200 W heating element for an environmental chamber is to be constructed from nickel wire of diameter 0·40 mm and resistivity $6·0 \times 10^{-7}$ Ω m. What length of wire is required if the element is to be connected to a 20 V_{rms} supply?

5.6   During an experiment it is found that a capacitor of 30 μF is required to make a low-pass filter for an electronic instrument. The only capacitors available have values 10, 15 and 20 μF. What series/parallel combination of these capacitors should be used?

5.7   A signal source E_B is connected to a measuring instrument of internal resistance 1·0 MΩ using a series coupling capacitor C, as shown in Fig. 5.10, in order to attenuate signals at the AC mains frequency (50 Hz). (a) What is the value of C which gives a lower cut-off frequency of 100 Hz? (b) What is the attenuation at 50 Hz?

5.8   A positive voltage step of 1·00 mV is applied at $t = 0$ to the input of an electrocardiograph. The variation of the output voltage with time is given below. What is the time constant τ' and the lower cut-off frequency of this instrument? (This response is the same as that of the CR circuit of Fig. 5.12c, for times between $t = 0$ and $t = T_1$.)

Time t/s	0	1	2	3	4	5
Voltage V_{out}/mV	1·00	0·57	0·35	0·22	0·13	0·09

5.9   If the voltage applied to the electrocardiograph of Problem 5.8 is removed at a time $t = 5s$, i.e. the input voltage drops to zero, describe the time variation of the output voltage and draw a sketch showing how the output voltage varies between $t = 0$ and $t = 10$ s.

5.10   Explain what is meant by an equivalent circuit.

5.11   The equivalent circuit for a short length of nerve fibre (Fig. 5.14) can be simplified to the RC circuit shown in Fig. 5.8 with $R = R_i R_m/(R_i + R_m)$ and $C = C_m$. If the time constant τ of the output signal is $2·0$ ms: (a) what is the value of C_m for a piece of fibre $2·0$ mm long; and (b) what is the capacitance per unit area of membrane if the fibre has radius 40 μm?

6

Biological Instrumentation and Measurement

Measurement systems are used to provide quantitative information about parameters that are of biological importance such as bioelectric potentials, muscle forces and blood pressure. A block diagram of a measurement system is shown in Fig. 6.1. The final stage of any measurement procedure requires a human observer to interpret the signal that is related to the biological activity. This signal may take many different forms, such as the movement of the mercury column in a thermometer or the movement of the hand of a stop watch. The most versatile signals are electrical voltages or currents because they can easily be amplified and modified to remove unwanted signals and they are easily displayed.

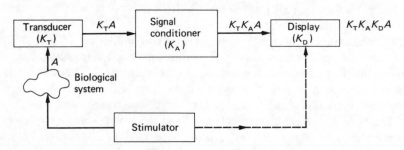

Figure 6.1 Schematic diagram of a biological measurement system.

A transducer is a device which converts any activity into an electrical signal. This signal is often small and contains additional information which is not related to the biological activity. The signal conditioner amplifies the voltage produced by the transducer and removes these unwanted signals. The conditioned signal is then fed to the display which presents the information to the observer. In some situations an external stimulus, such as an electrical pulse, is required to trigger the biological activity. This is provided by the stimulator which may also be used to trigger the display.

Each of the components in the system has a transfer function, K, which relates the output signal to the input signal according to OUTPUT $= K$(INPUT). These functions must be known if a quantitative value of the activity is to be determined. Consider, for example, the measurement of the movement of a limb, the output signal being drawn on graph paper. The transfer functions of the components are K_T (V mm^{-1}), K_A and K_D (mm display V^{-1}) as shown in

Fig. 6.1. The activity A is detected by the transducer which produces a voltage $K_T A$ at its output. (Note that the units are correct.) The voltage at the output of the conditioner is $K_T K_A A$ and the output D (mm) of the display is $K_T K_A K_D A$ (mm of display). The activity is related to the displayed quantity by

$$A = \frac{D}{K_T K_A K_D}$$

If $K_T = 0 \cdot 1$ V mm^{-1}, $K_A = 100$ and $K_D = 100$ mm V^{-1} one millimetre movement of the display corresponds to 1×10^{-2} mm movement of the limb. In order that no distortion is introduced into the signal the frequency response and the time constant (§ 5.3) of the measurement system must be appropriate to the signal that is being monitored. This will be discussed in more detail when the different types of transducer and measurement system components are described.

6.1 Electrodes

The purpose of an electrode is to convert an ionic bioelectric potential into an electronic current or voltage which can be amplified and measured. There are three basic types of electrode: microelectrodes which are used to detect potentials within a single cell; needle electrodes which are used to measure the activity in a localized region of tissue; and surface electrodes which are used to measure the potentials at the surface of a system. Examples of these types of electrode are shown in Fig. 6.2 together with the potentials that they detect and typical dimensions.

The measurement of the action potentials of individual cells is difficult because of the small size of the cell. Typically nerve cell diameters range from 1 to 650 μm. For intracellular measurement the electrode diameter should be at least ten times smaller than the smallest dimension of the cell. A micropipette electrode (Fig. 6.2a) consists of a glass capillary drawn down to a narrow tapered tip. The bore of the capillary is filled with an electrolyte solution, often KCl, and the electrical connection made using wire, such as silver–silver chloride. The action potential due to the firing of a single neuron is shown in Fig. 6.2b. The potential spikes caused by a neuron are of constant amplitude. When the action potential of a large number of neurons is measured using a needle electrode, or a small surface electrode, the signal is the summation of a number of individual action potentials that occur at random (Fig. 6.2d). A floating type of surface electrode (Fig. 6.2e) consists of a metal disc mounted in a plastic holder. The space between the disc and the skin is filled with an electrolyte paste, such as a sodium chloride–glycerol mixture. Some of the electrophysiological signals that can be detected using surface electrodes are indicated in Table 6.1 together with the origin of the activity and the typical range in size and frequency content of the signals.

The biopotentials are determined by measuring the potential differences between the output terminals of two electrodes located in different parts of the biological system. The transfer function is governed by the electrical properties of the electrodes. Although the different types of electrode vary in detail the principle of operation is the same. A micropipette and reference electrode used

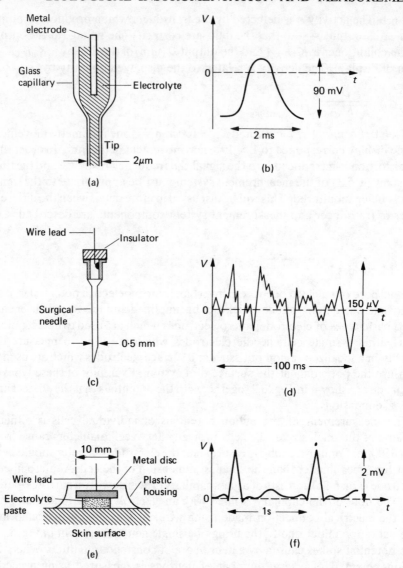

Figure 6.2 (a) Micropipette electrode. (b) Action potential of a single neuron. (c) Needle electrode. (d) Electromyogram (EMG). (e) Surface electrode (floating type). (f) Electrocardiogram (ECG).

Table 6.1 Some electrophysiological parameters.

	Origin	Signal size	Frequency range/Hz
Electrocardiogram (ECG)	Heart muscle	0·1–4 mV	0·05–100
Electromyogram (EMG)	Skeletal muscle	0·05–1 mV	10–3000
Electroencephalogram (EEG)	Brain	10–100 μV	0·05–100

for intracellular measurement are shown in Fig. 6.3a. With all electrodes there is a metal–electrolyte interface. Ions from the metal migrate into the electrolyte solution and ions from the solution combine with the metal. The charges carried by these ions are of opposite sign so that an electrical double layer is formed and a potential difference is developed across the interface. This is called the electrode potential. It is not related to the biological activity. Since electrodes are used in pairs, the electrode potentials should cancel when the potential difference between the output terminals is measured. However, small random fluctuations in electrode potential occur which do not cancel and cause a drift in the output voltage.

If the electrode potential is ignored an electrode can be represented by the equivalent circuit shown in Fig. 6.3b where R_E represents the electrode resistance and C_E the electrode capacitance. In the micropipette electrode most of the resistance is due to the electrolyte solution in the tip (see Example 5.1). The walls of the electrode are of glass, which is a good insulator. Since the insulator separates two electrical conductors the arrangement behaves like a capacitor, C_E. For micropipette electrodes values of R_E range from about 10 to 200 MΩ and C_E from 2 to 10 pF.

The RC circuit of Fig. 6.3b is the same as the low-pass filter of Fig. 5.8. The transfer function $K = V_{out}/V_{in} = V_{out}/E_B$ is, from Equ. 5.10,

$$\frac{V_{out}}{E_B} = \frac{1}{\sqrt{[1 + (2\pi f R_E C_E)^2]}} \tag{6.1}$$

so that the transfer function decreases with increasing frequency, i.e. high-frequency signals are attenuated more than low-frequency signals.

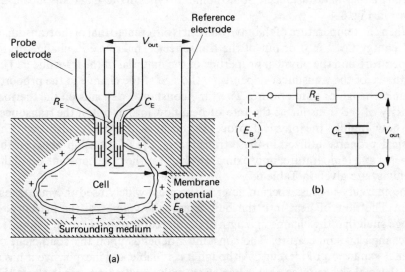

(a)

(b)

Figure 6.3 (a) A pair of electrodes used for intracellular measurement. (The symbols R_E and C_E indicate the origin of the electrode resistance and capacitance. They are not a resistor and capacitor connected to the electrode.) (b) The equivalent circuit.

Example 6.1

An electrode with $R_E = 50$ MΩ is to be used to detect bioelectric signals which have components at 100 Hz. What is the maximum value that the electrode capacitance can have such that the transfer function is not less than 0·50 at this frequency, i.e. the attenuation, E_B/V_{out}, is not greater than 2·0?

From Equ. 6.1, $C_E^2 = [(E_B/V_{out})^2 - 1]/(2\pi f R_E)^2$ so that

$$C_E^2 = \frac{(4\cdot0 - 1)}{(2 \times 3\cdot14 \times 100 \text{ s}^{-1} \times 50 \times 10^6 \text{ }\Omega)^2}$$

$$C_E = 55 \text{ pF}$$

Many bioelectric signals are sharp pulses or spikes of short duration (see Figs. 6.2b and d). The presence of the electrode capacitance C_E will cause the signal at the output of the electrodes to be distorted, as shown in Fig. 5.12. In order to obtain good reproduction of the potential spike the time constant $\tau = R_E C_E$ of the electrode (see § 5.3) should be much smaller than the pulse duration T_1. Usually τ equal or less than $T_1/10$ is taken as a suitable criterion. For the electrode of Example 6.1 which has $R_E = 50$ MΩ and $C_E = 55$ pF the time constant $\tau = 2\cdot8$ ms so that pulses with duration less than about 28 ms will be distorted.

6.2   Thermal transducers

Many of the processes that occur in living organisms are dependent upon temperature so that the determination of temperature is one of the most important physiological measurements. The temperature of a system is measured using a thermal transducer. Some common types of thermal transducers are shown in Fig. 6.4.

When the temperature of the environment is different to that of the transducer heat energy flows into or out of the transducer. This causes a change in the temperature and the physical properties of the material of the transducer. The sensitivity of the transducer depends on the size of the change in the property for unit change of temperature. The time constant τ depends on the thermal capacity of the transducer, the rate of heat flow into or out of the transducer and how rapidly the physical processes occur within the transducer. The physical property utilized in these transducers, the relationship between the property and temperature, approximate values for the time constant τ and the sensitivity are given in Table 6.2.

The mercury-in-glass thermometer is the most widely used thermal transducer. The flow of heat into the bulb causes the volume of the mercury to change such that the change in length of the mercury column is proportional to the change in temperature. The constant α depends upon the coefficient of thermal expansion of mercury. Although it is reliable and inexpensive it has a high thermal capacity so that when used with small organisms it alters the temperature which is being measured and it has a long time constant so that rapid changes in temperature cannot be measured. It does not produce an electrical output and cannot be used for continuous monitoring.

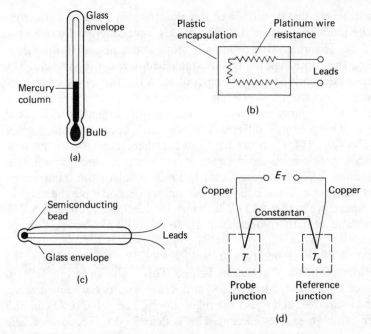

Figure 6.4 Some common types of thermal transducers. (a) Mercury-in-glass thermometer. (b) Platinum resistance thermometer. (c) Thermistor. (d) Thermocouple.

Thermistor and thermocouple transducers are widely used in biological measurement because they do not have these disadvantages. A thermistor consists of a bead of semiconducting material enclosed in a thin glass envelope which is filled with an inert gas. Electrically a semiconductor is intermediate between a metallic conductor and an insulator. The origin of the charge carriers in a semiconductor is different to that in a metal and its resistivity decreases with increasing temperature (see § 5.1). The relationship between the resistance R_T of a thermistor and the absolute temperature $T(K)$, see Chapter 9, is given in Table 6.2 where $R_0(\Omega)$ is the resistance at a reference temperature $T_0(K)$. The constant B depends upon the material and typically has a value of about 3×10^3 K. The main disadvantage of a thermistor is the non-linear relationship between resistance and temperature.

Table 6.2   The thermal transducers of Fig. 6.4.

	Property	Equation	τ/s	Sensitivity
(a)	Thermal expansion	$L_T = L_0[1 + \alpha(T - T_0)]$	10	High
(b)	Electrical resistance	$R_T = R_0[1 + \beta(T - T_0)]$	1	Low
(c)	Electrical resistance	$R_T = R_0\, e^{B(1/T - 1/T_0)}$	0·5	High
(d)	Contact potential	$E_T = A\Delta T + B\Delta T^2$	0·05	Low

When two different metals are joined together to form a junction electrons flow across the junction because different metals have different free electron densities (§ 5.1). This migration eventually stops and a potential difference, which is called the contact potential, is established across the junction. The contact potential depends upon the temperature. A thermocouple consists of two junctions, a reference junction which is held at a fixed reference temperature, T_0, and a probe junction which is placed on the organism which is at temperature T. The potential difference E_T depends upon the temperature difference $\Delta T = (T - T_0)$ as indicated in Table 6.2 where A and B are constants. If the reference junction is placed in an ice–water mixture, which by definition has a temperature of 0.00 degree Celsius, ΔT is equal to the temperature (degree C) of the probe. For a copper–constantan thermocouple the value of A at room temperature is about 40 μV K^{-1} and B is small, see Problem 6.5. The main disadvantage of a thermocouple is its low sensitivity, but this can be overcome using a high amplification signal conditioner.

The resistance of resistance type transducers can be measured using an electrical network called a Wheatstone bridge. This is shown in Fig. 6.5a. R_1 is the resistance of the transducer, R_2, R_3 and R_4 are resistors of known value, one of which is variable, and E_b is the bridge supply. The detector D can be a galvanometer, which measures current. This is described in Section 6.5. The voltage V_a at a equals the voltage at c minus the voltage drop $I_1 R_1$ across R_1, $V_a = V_c - I_1 R_1$. Similarly the voltage at b is $V_b = V_c - I_2 R_2$. The potential difference, $V_{ab} = V_a - V_b$, across ab is

$$V_{ab} = I_2 R_2 - I_1 R_1 \qquad (6.2)$$

From Ohm's law the current I_1 is the potential difference E_b divided by the total resistance $(R_1 + R_3)$, i.e. $I_1 = E_b/(R_1 + R_3)$ and similarly $I_2 = E_b/(R_2 + R_4)$. Substituting into Equ. 6.2 gives

$$V_{ab} = E_b \left(\frac{R_2}{R_2 + R_4} - \frac{R_1}{R_1 + R_3} \right) \qquad (6.3)$$

One of the resistors, say R_2, can be varied until $V_{ab} = 0$ so that no current flows

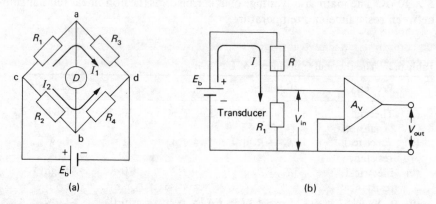

(a)   (b)

Figure 6.5 (a) A Wheatstone bridge for measuring the resistance of a transducer. (b) A circuit for measuring rapid changes in resistance.

through the galvanometer. This is called the balance condition. Balance or null conditions are often used in measurement systems. When the bridge is balanced $R_2/(R_2 + R_4) = R_1/(R_1 + R_3)$ or

$$R_1 = (R_3/R_4)R_2 \qquad (6.4)$$

If the ratio (R_3/R_4) is made unity the resistance R_1 of the transducer is equal to R_2. This arrangement can be used to measure steady or slowly changing temperatures.

When the temperature changes are rapid, as in the muscle twitch experiment, a circuit similar to that shown in Fig. 6.5b can be used. R is a fixed resistor and A_V is a voltage amplifier (see § 6.4). From Ohm's law the current I is the supply voltage E_b divided by the total resistance $(R + R_1)$ where R_1 is the resistance of the transducer, $I = E_b/(R + R_1)$. The voltage V_{in} across R_1, which is the voltage applied to the input of the amplifier, is $V_{in} = IR_1 = E_b R_1/(R_1 + R)$. If R is made much larger than R_1 the resistance R_1 can be ignored with respect to R so that $V_{in} = E_b R_1/R$. The voltage V_{out} at the output of the amplifier is $V_{out} = A_V V_{in}$ so that

$$V_{out} = \frac{A_V E_b}{R} R_1$$

The output voltage is proportional to the thermistor resistance which is related to the temperature. Electronic thermometers are available, based upon thermistor or thermocouple transducers, which give a direct reading of temperature on a calibrated scale.

6.3 Displacement, force and acceleration transducers

Mechanical parameters such as relative movement, deformation and force have to be measured over a wide range of conditions. For example in muscle physiology two extreme situations occur. Forces developed by muscle tissue are measured when the muscle is held at constant length (isometric measurement) or the change in the length of the muscle at constant force is measured (isotonic measurement). The transducer used to measure a biomechanical response must be chosen to suit the needs of the particular application.

Three types of transducer used for the measurement of displacement or deformation are shown in Fig. 6.6. The simplest is the variable-resistance (or potentiometer) transducer which is used to measure relatively large movements. It consists of a wire resistance coil with a sliding contact, the slider being attached to the moving member. The electrical resistance between the output terminals is proportional to the position of the slider. These transducers have high sensitivity but their time constant τ is long (about 0.3 s) and they do not have good resolution because of the discontinuous nature of the wire coil.

Small deformations of a structure, i.e. the elongation or the compression of the material of the structure, such as cardiac muscle fibres, can be measured using a strain gauge. This is shown in Fig. 6.6b and consists of a piece of high-resistance wire mounted on a flexible plastic substrate. The gauge is attached to the structure under investigation. When the structure elongates, the length

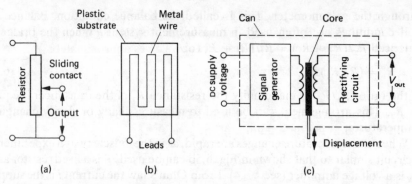

Figure 6.6 (a) A potentiometer displacement transducer. (b) A strain gauge. (c) A linear voltage displacement transducer (LVDT).

L of the wire increases and the cross-sectional area A decreases so that the resistance increases (see Equ. 5.3). The change in resistance is proportional to the elongation. The sensitivity of a strain gauge is described by the gauge factor G which is defined as the ratio of the fractional change in the resistance, $\Delta R/R_0$, to the fractional change in length $\Delta L/L_0$,

$$G = \frac{\Delta R/R_0}{\Delta L/L_0} \tag{6.5}$$

Gauge factors range from about 1 to 120. When used in conjunction with the appropriate conditioner and display, strain gauges have high sensitivity and small time constant.

Strain gauges are often used in pairs in order to increase the sensitivity and to compensate for changes in temperature (which cause a change in the resistance). A typical application is shown in Fig. 6.7. The active gauges form two arms of a Wheatstone bridge (Fig. 6.5a) the other arms being identical strain gauges which do not deform and have a fixed resistance R_0. These are called dummy gauges. When the bone bends, as shown, the upper gauge elongates and has resistance $(R_0 + \Delta R)$ and the lower gauge is compressed and has resistance $(R_0 - \Delta R)$. R_0 is the resistance of the undeformed gauge. The

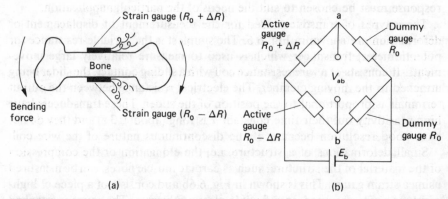

Figure 6.7 (a) Placement of the strain gauges to measure the deformation of bone. (b) The Wheatstone bridge for measuring ΔR.

output voltage V_{ab} of the bridge is given by Equ. 6.3 with $R_1 = R_0 + \Delta R$, $R_2 = R_0 - \Delta R$ and $R_3 = R_4 = R_0$. Re-arranging and setting ΔR much less than R_0 (so that ΔR can be ignored with respect to R_0) gives approximately

$$V_{ab} = -\frac{E_b}{2R_0}\Delta R \tag{6.6}$$

The bridge output voltage is proportional to the change in resistance which is proportional to the deformation of the structure.

Example 6.2

A pair of strain gauges, with $G = 2 \cdot 0$ and $R_0 = 120\ \Omega$, are used to measure the bending deformation of a bone. If the bridge supply voltage $E_b = 10$ V what is the fractional deformation $\Delta L/L_0$ at the surface of the bone if the bridge output voltage $V_{ab} = -10$ mV.

From equations 6.5 and 6.6 the fractional change in length is $\Delta L/L_0 = \Delta R/R_0 G = 2V_{ab}R_0/E_b R_0 G = 2V_{ab}/E_b G$. Substituting values gives

$$\frac{\Delta L}{L_0} = \frac{2 \times 10 \times 10^{-3}\ \text{V}}{10\ \text{V} \times 2\cdot0} = 1 \times 10^{-3} = 0\cdot1\%$$

The linear voltage displacement transducer (LVDT) shown in Fig. 6.6c has a better resolution than the potentiometer transducer and can be used to measure a wide range of displacements. It consists of two coils of wire, the primary and the secondary, wound one upon the other. An alternating current, from a signal generator, is fed through the primary so that an alternating magnetic field is produced. This field passes through the secondary coil and induces a potential difference across its ends. This process is called mutual induction and is similar to self-induction described in Section 5.2. The alternating induced voltage is converted to a direct (DC) voltage by a circuit called a rectifier. The amplitude of the induced voltage depends on the position of the core, which can be moved along the axis of the coils. The DC output voltage is directly proportional to the position of the core. The coils, signal generator and electronic circuit are fitted inside a metal can. These transducers have a high sensitivity and a time constant of about 0·2 s.

The most widely used force and pressure transducers consist of a high stiffness elastic member, to which the force is applied, and a displacement transducer which measures the deformation of the member. A simple cantilever transducer with two bonded strain gauges is shown in Fig. 6.8a. The deformation $\Delta L/L_0$ of the cantilever, and hence the change in resistance, $\Delta R/R_0$, of the strain gauges is proportional to the applied force F. $\Delta R/R_0$ is measured using a Wheatstone bridge. The transducer is calibrated by attaching known masses to the cantilever, which produce a force $F = Mg$, and noting the bridge output voltage V_{ab}. A pressure transducer that operates in the same way is shown in Fig. 6.8b. The fluid to be measured is taken from the system (such as blood from an artery) through a chamber and then back into the system. One of the walls of the chamber is a flexible diaphragm. The deformation of the diaphragm, which is measured using an LVDT, is proportional to the fluid pressure. These transducers are useful for measuring steady or slowly varying forces and pressures.

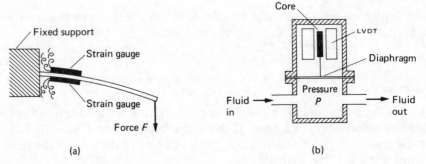

Figure 6.8 (a) A cantilever force transducer. (b) A pressure transducer.

Rapidly changing forces can be measured using a piezoelectric transducer. When certain materials, such as quartz and barium titanate, are deformed electrical charges appear on the surface of the material. This is called the piezoelectric effect. A piezoelectric force transducer is shown in Fig. 6.9a. The charges collect on metal electrodes deposited onto the surfaces of the piezoelectric disc. These electrodes are connected to a special type of amplifier, called a charge amplifier, which produces a voltage V_{out} at its output which is proportional to the charge Q at its input. Since Q is proportional to the deformation $\Delta L/L_0$ of the disc, which is proportional to the applied force F, the output voltage is directly related to F. These transducers can be used to measure alternating and impulsive forces but are not suitable for measuring steady forces. Since their stiffness is very high they can be used for isometric measurement.

A piezoelectric accelerometer which operates in the same way is shown in Fig. 6.9b. A mass M is mounted on top of the piezoelectric disc. When the

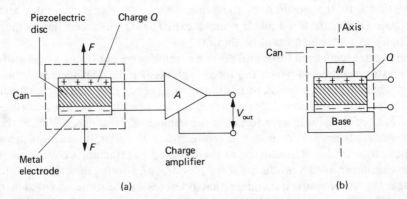

Figure 6.9 (a) A piezoelectric force transducer with charge amplifier. (b) A piezoelectric accelerometer.

accelerometer moves up and down along its axis with acceleration a the mass exerts a force $F = Ma$ on the disc so that the output voltage V_{out} of the charge amplifier is proportional to the acceleration. Accelerometers with a small total mass (about 0·5 g) are available which are suitable for biomechanical measurement on man and animals.

6.4 Signal amplification and conditioning

The purpose of the signal conditioner is to amplify the small signals derived from the transducer and to remove unwanted signals. The principle of signal amplification can be explained by considering a hydraulic flow system such as that shown in Fig. 6.10a. Liquid from a reservoir flows down a pipe which contains a valve. The flow rate Q (m³s⁻¹) can be controlled by the valve, small changes in valve position causing large changes in the flow rate. The change in valve position can be considered as the input signal and the change in the flow rate as the output signal.

A transistor operates in a similar way. A bipolar transistor consists of three slabs of semiconducting material (Fig. 6.10b) called the emitter (E), base (B) and collector (C). When a voltage V is established across the collector–emitter terminals a large current I_C flows into the collector and out of the emitter. This current is controlled by the current I_B flowing into the base. The base acts in the same way as the valve in the hydraulic analogy, small changes ΔI_B in the base current causing large changes ΔI_C in the collector current. Thus there is current amplification. The current gain A_I is defined as $A_I = \Delta I_C/\Delta I_B$ where ΔI_B is the input signal and ΔI_C the output signal.

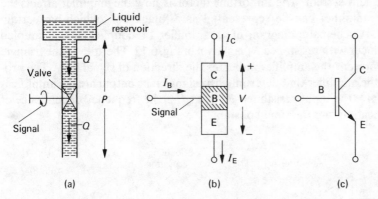

(a)   (b)   (c)

Figure 6.10 (a) A hydraulic amplifier. (b) A schematic diagram of a bipolar transistor. (c) The symbol for a transistor.

The circuit for a simple transistor amplifier is shown in Fig. 6.11a. E_b is a DC voltage supply which provides the energy for the circuit to operate. It is equivalent to the liquid reservoir in the hydraulic analogue. A resistor R_L is placed in the collector lead. When the collector current changes by an amount ΔI_C, due to a small change ΔI_B, the voltage across R_L changes by an amount $\Delta V_L = R_L \Delta I_C$. The base signal current ΔI_B can be produced by applying a voltage input signal ΔV_B across the base–emitter terminals. A small voltage change ΔV_{B} at the input will cause a large voltage change ΔV_{L} at the output so that there is voltage amplification. The voltage gain A_{V} is defined as $A_{\mathrm{V}} = \Delta V_{\mathrm{L}}/\Delta V_{\mathrm{B}}$. We will call ΔV_{B} the input signal V_{in} and ΔV_{L} the output signal V_{out}.

The current and voltage amplifications produced by a single amplifier stage are typically 200 and 100 respectively. Often this is insufficient and it is necessary to arrange several stages in series. An integrated circuit (IC) consists of a number of amplifying stages constructed on the same slab of semiconducting material. The physical appearance of two types of IC is shown in Fig. 6.11b. The inherent amplification A_{V} is of the order of 10^5. The IC is used with various combinations of external components (resistors, capacitors and inductors) to produce an amplifier with the desired properties.

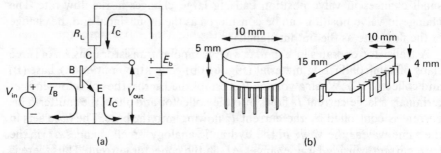

Figure 6.11   (a) A single stage transistor voltage amplifier. (b) Two integrated circuits.

The way in which an amplifier operates is not important in the analysis of a measurement system. The important factor is how the amplifier affects the signal. An amplifier can be represented as a 'black box' which has certain properties. A schematic diagram of an amplifier DC coupled and AC coupled to a transducer with resistance R_{E} is shown in Fig. 6.12. The apex of the triangle used to represent the amplifier indicates the direction of signal flow. The properties of the amplifier that determine the signal at its output are the amplification A_I or A_{V}, the input resistance $R_{\mathrm{in}} = V_{\mathrm{in}}/I_{\mathrm{in}}$, the frequency dependence of the amplification and the time constant.

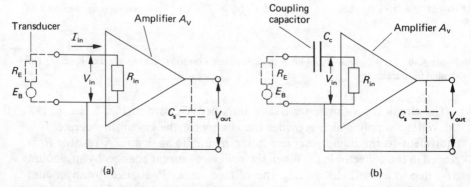

Figure 6.12   Schematic diagram of an amplifier with input resistance R_{in} and stray capacitance C_{s}: (a) DC coupled, and (b) AC coupled, to a transducer.

Example 6.3

The voltage E_B from a pair of electrodes with resistance $R_E = 10 \, M\Omega$ is applied to the input of an amplifier which has input resistance $R_{in} = 10 \, M\Omega$ as shown in Fig. 6.12a. If the amplification $A_V = 1.0 \times 10^3$ and the biopotential $E_B = 1.0 \, mV$ what is the voltage at the output of the amplifier? (Ignore the stray capacitance C_s.)

The output voltage $V_{out} = V_{in}A_V$ where V_{in} is the voltage at the input of the amplifier, which does not equal E_B. From Ohm's law $V_{in} = I_{in}R_{in}$ where, again from Ohm's law, $I_{in} = E_B/(R_E + R_{in})$ so that $V_{out} = E_B A_V R_{in}/(R_E + R_{in})$. Substituting values

$$V_{out} = \frac{1.0 \times 10^{-3}V \times 1.0 \times 10^3 \times 10 \times 10^6 \Omega}{10 \times 10^6 \Omega + 10 \times 10^6 \Omega} = 0.5 \, V$$

If $R_{in} = 100 \, M\Omega$, $V_{out} = 0.99 \, V$. In order to obtain the maximum output voltage the input resistance R_{in} of the amplifier must be much larger than the resistance R_E of the transducer.

An amplifier consists of an arrangement of active devices (transistors) and passive circuit elements (resistors and capacitors). Since the impedance of these elements depends upon the frequency of the signal (§ 5.2) the amplification depends upon the frequency. The amplifier can be connected to the transducer in two ways. It can be directly coupled (DC coupled), Fig. 6.12a, or it can be coupled using a series capacitor C_c (AC coupled), Fig. 6.12b. With DC coupling all of the signals from the transducer are presented to the amplifier whereas with AC coupling only the alternating signals appear at the input because C_c has infinite impedance for DC signals (see § 5.2). The amplifying stages in an amplifier can also be connected directly together, DC coupled, or with a coupling capacitor, AC coupled. The frequency dependence of the amplification of a DC and an AC coupled amplifier is shown in Fig. 6.13. The gain of the AC coupled amplifier is zero at low frequencies because of the infinite impedance of C_c. The amplification of both types of amplifier decreases at high frequency because of stray capacitance, C_s, which 'shunts' the signal. The operating frequency range of an amplifier is described by the upper and lower cut-off frequencies f_u and f_l. These are the frequencies at which the

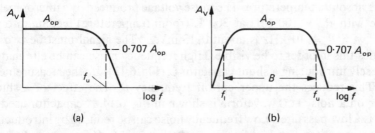

Figure 6.13 The frequency dependence of the amplification for: (a) a DC coupled amplifier, and (b) an AC coupled amplifier.

amplification is 0·707 times that in the plateau region. The bandwidth B of an amplifier is the frequency range between f_u and f_1, i.e. $B = f_u - f_1$. The amplification in the normal operating range of an amplifier is indicated as A_{op}.

Example 6.4

An AC coupled amplifier with $A_{op} = 1 \times 10^3$, $f_1 = 10$ Hz and $B = 3·3$ kHz is used to amplify an EMG signal which has components between 30 Hz and 4 kHz. Describe the composition of the signal at the output of the amplifier.

Since $f_1 = 10$ Hz and $B = 3·3$ kHz, f_u is approximately 3·3 kHz. The amplification is approximately equal to A_{op} for signals with frequencies greater than about $3 f_1$ and less than $0·3 f_u$. For this example, EMG signals in the range 30 to about 1 kHz will be amplified by a factor of 1×10^3. At higher frequencies the amplification will be less, and the signal at the output of the amplifier will emphasize the low frequency components. Thus the output voltage is not a faithful reproduction of the input.

Biomeasurement systems must be capable of reproducing pulses as well as DC and alternating signals. Good reproduction of a pulse requires DC coupling and an infinite bandwidth. In many cases it is necessary to employ AC coupling (to block unwanted changes in DC level). The coupling capacitor C_c introduces a time constant τ' into the measurement system (see § 5.3). The value of the time constant, $\tau' = R_{in} C_c$, should be large compared to the duration of the pulse. If τ' is about the same as the pulse duration the output voltage will be distorted as shown in Fig. 5.12c.

Spurious information is introduced into the signal by the transducer, through the connecting leads and by the amplifier. This is called noise. There are basically two types of noise: low frequency noise which is caused by the pick-up of low frequency and mains (50 Hz) voltages in the biosystem and connecting leads, and high frequency noise which is generated in the transistors and resistors. Since many of the signals of biological origin are very small it is necessary to reduce the noise to a minimum. Pick-up can be avoided by placing the biosystem and the signal conditioner in earthed metal containers. This is called shielding. Shielded connecting leads should also be used. High frequency noise is due to the random fluctuation of the flow of electrons. The RMS noise voltage V_{rms} generated in a resistor R feeding into an amplifier with bandwidth B is $V_{rms} = \sqrt{4kTBR}$ where k is Boltzmann's constant ($= 1·38 \times 10^{-23}$ J K^{-1}) and T the absolute temperature. The noise voltage produced by a micropipette electrode with $R_E = 100$ MΩ at 300 K (room temperature) feeding into an amplifier with $B = 10$ kHz is about $0·13$ mV$_{rms}$. The signal must be two or three times this in order to be visible. High frequency noise can be eliminated by purposely introducing a shunt capacitor C_s (Fig. 6.12) into the measurement system. This reduces the upper cut-off frequency f_u. The effect of a shunt capacitor on a noisy ECG waveform is shown in Fig. 6.14. A capacitor used in this way is a low-pass filter. Low frequency noise can be reduced by introducing a coupling capacitor C_c which increases the lower cut-off frequency f_1. Both of these procedures introduce undesired time constants into the system.

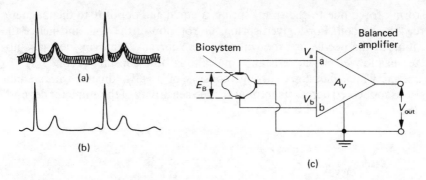

Figure 6.14 A noisy ECG waveform. (b) Elimination of noise using a shunt capacitor. (c) A balanced amplifier.

Low frequency noise can be eliminated without affecting the low frequency signals by using a balanced amplifier. This type of amplifier, which is widely used in biomeasurement systems is shown in Fig. 6.14c. It consists of an amplifier with two live inputs a and b instead of one live and one earthed input as in the amplifier shown in Fig. 6.12. The output V_{out} of the amplifier is given by $V_{out} = A_V(V_a - V_b)$ where A_V is the amplification. If the centre of the biosystem is fixed at earth potential (0 V) by a third electrode the biopotential E_B will cause the potential on the upper electrode to rise to $E_B/2$ and that on the lower electrode to fall to $-E_B/2$. The noise potential E_N picked-up by the leads is the same for each input and adds to the biosignal. Thus $V_a = E_B/2 + E_N$ and $V_b = -E_B/2 + E_N$ so that $V_{out} = A_V E_B$. The noise signal has been eliminated.

6.5 Signal display

The signal display provides the observer with a quantitative indication of the biological activity (provided no distortion has been introduced by the other components). Signals can be displayed in two ways, in analogue form and in digital form. The analogue display gives a continuous reading of the measured variable on a meter, calibrated paper or screen. With digital display the variable is indicated as a set of digits on light-up electronic number tubes or printed on paper tape. A digital meter samples the signal periodically and does not give a continuous reading of the variable being measured.

Most meter movements, i.e. those with an indicating needle and calibrated scale, are based upon the moving coil galvanometer (Fig. 6.15). This consists of a small coil of wire mounted on frictionless bearings between the poles, N and S, of a permanent magnet. A needle is attached to the coil. When a current I passes through the coil the electrons which comprise the current experience a force because of the magnetic field (§1.4). Since the electrons are constrained to move inside the wire this force is transmitted to the coil. For the coil in the position shown there is no force on the upper and lower sides because the direction of current is parallel to the magnetic flux (from N to S). For the vertical sides, of length L, the direction of current flow is perpendicular to B and they experience a magnetic force of magnitude $|F_M| = ILB$. These forces are in opposite directions, as shown in Fig. 6.15, and they create a torque which causes the coil to rotate. This rotation continues until the restoring force due to the spiral spring is equal and opposite to the magnetic force on the coil. For a given spring the rotation of the coil, and hence the position of the needle, is proportional to the current. A moving coil galvanometer has low electrical resistance and a long time constant (about 1 s). An ammeter, which measures current, consists of a galvanometer with a shunt resistor connected across its terminals. The sensitivity of the ammeter depends on the value of this resistor.

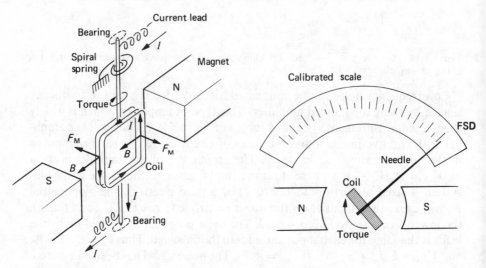

Figure 6.15   A moving coil galvanometer.

A galvanometer can be modified so that it will measure voltage by introducing a resistance R_M in series with the meter movement as shown in Fig. 6.16a. The resistance of the movement is small so that the current I through the galvanometer is $I = V/R_M$ where V is the voltage applied to the input terminals of the voltmeter. Since the position of the needle is proportional to I this system can be used to measure voltage. The way in which an ammeter and a voltmeter are used is shown in Fig. 6.16b. The ammeter measures the current through the circuit element Z and the voltmeter measures the potential difference (i.e. voltage) across Z. The instruments should have no effect on the signal in the circuit so that an ammeter must have very low resistance and a voltmeter must have very high resistance. This high resistance cannot always be obtained using a passive circuit element, it is necessary to replace R_M (Fig. 6.16a) by a high input resistance amplifier. These are called electronic voltmeters and have input resistances of several thousand megohms.

A meter movement can be adapted to measure alternating currents and voltages. This is done by converting the AC signal into a DC signal using a rectifying circuit. Most AC meters are calibrated to read the RMS value of a sinusoidal signal (V_{rms} = amplitude/$\sqrt{2}$). AC meters only operate over a given frequency range, typically 15 Hz to 5 MHz. Their time constant is governed by the response of the meter movement (of the order of seconds) so that they cannot be used to measure rapid changes in the amplitude of an AC signal.

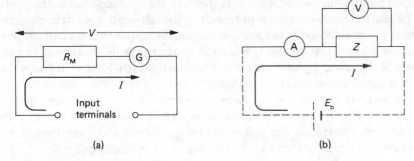

Figure 6.16 (a) A voltmeter. (b) The use of an ammeter and a voltmeter.

The main disadvantages of meter and digital display systems are their long time constants and the fact that they do not give a permanent record of the time variation of the signal. This can be obtained using a chart recorder which employs a writing device to draw the signal on calibrated paper. Two types of recorder based on galvanometer movements are shown in Fig. 6.17. The parameters that govern their usefulness for data retrieval are the sensitivity (mm display movement for unit input signal), the paper speed (mm s^{-1}) and the maximum frequency that they can write. For the pen galvanometer recorder the maximum frequency is about 50 Hz and for the ultraviolet recorder it is about 10 kHz.

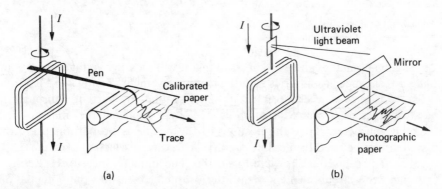

Figure 6.17 (a) A galvanometer pen recorder. (b) An ultraviolet light recorder.

The most versatile display system is the oscilloscope (Fig. 6.18), which uses an electron beam to draw the signal on a fluorescent screen. The source of electrons is called the electron gun. When the filament is heated, by passing a current through it, electrons are given off. These negatively charged electrons are accelerated away from the filament because of the large (positive) potential difference between the anode and the filament. When the electrons reach the anode they pass through the hole in the centre, travel down the tube and strike the screen. This causes the screen to glow and a small point of light is observed. The brightness of the spot is controlled by a negative potential (which repels the electrons) applied to the grid electrode. The shape of the spot on the screen is controlled by the focussing anode. The position of the spot can be changed by applying a voltage across the X or the Y plates. When the right-hand X plate is made positive with respect to the left-hand X plate the electron beam is attracted towards the positive plate and the spot on the screen moves to the right (see §1.4). Similarly when the upper Y plate is made positive with respect to the lower Y plate the spot is deflected upwards. If the voltages on the plates are constant the spot is stationary.

Oscilloscopes are complex measuring systems which have a large number of controls. The simplest way to learn how to operate the instrument is to determine experimentally the function of these controls. If possible the following discussion should be read whilst looking at an oscilloscope.

In the most usual mode of operation a continuously increasing voltage V_X is applied to the X plates (Fig. 6.18b), so that the electron beam is swept across the screen (from left to right looking onto the screen). V_X is produced internally by the timebase generator circuit. The rate of sweep is set by the timebase control which is calibrated in time per unit X deflection. A timebase setting of 1 ms/cm means that 1 cm of trace corresponds to 1 ms in time. When the beam reaches the right-hand side of the screen it is returned very rapidly (so that it cannot be seen) to the left-hand side and the sweep occurs again.

The signal to be measured is connected to the input terminals of the oscilloscope, it is amplified and then applied to the Y plates of the CRT. This causes a deflection in the y direction and a trace is formed on the screen which shows the time variation of the signal. The amplifier is calibrated in volts per division so that a particular Y deflection corresponds to a certain voltage at the input.

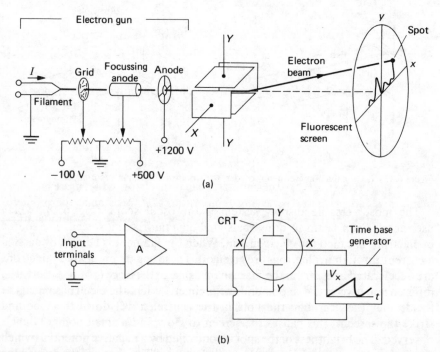

Figure 6.18 (a) A cathode ray tube (CRT). The entire system is enclosed in an evacuated glass envelope. (b) An oscilloscope block diagram.

A 2·5 cm deflection with the sensitivity control at 0·1 V/cm corresponds to a voltage of 0·25 V at the input. With many oscilloscopes two traces can be displayed simultaneously so that two signals can be compared. In order to obtain a stationary trace of a periodic waveform the timebase sweep must occur at the correct time. This is done by internal triggering circuits which control the timebase generator. If the trigger controls are not set correctly the trace moves slowly across the screen.

When examining a single event it is necessary to arrange that the sweep occurs only once. This is done by operating the oscilloscope in the single-shot mode. The sweep can be triggered manually or from a signal derived from the stimulator which is used to initiate the biological activity. It is possible, using this type of arrangement to measure the time period between the stimulation and the response of the biosystem. With many oscilloscopes the glow time of the screen is short so that a complete trace of a slow-moving event cannot be seen. A storage oscilloscope will hold the trace on the screen for an indefinite period of time. When used in the single-shot mode a storage oscilloscope is useful for measuring low frequency periodic signals (e.g. ECG) and random signals (e.g. EMG).

Oscilloscopes typically have a maximum sensitivity of about 1 mV/cm deflection and an input resistance of 1 MΩ. The time constant, τ, is less than 1 μs so that they can be used to measure steady or very rapidly changing signals. However, oscilloscopes do not provide a permanent record of the signal and it is difficult to obtain an accurate value of a voltage.

Problems

6.1 An electrode has resistance $R_E = 100$ MΩ and capacitance $C_E = 10$ pF. What is the frequency at which the attenuation is 2·0?

6.2 A range of electrodes are available which have the same capacitance but different resistance. If the value of C_E is 10 pF determine the resistance R_E of an electrode that is suitable for examining the action potential of a squid giant axon for which the pulse duration T_1 is 2·0 ms. Assume that $\tau = T_1/20$.

6.3 Compare the advantages and disadvantages of using a mercury-in-glass thermometer and a thermistor to measure temperature.

6.4 A platinum resistance thermometer can be used for the accurate measurement of steady or slowly varying temperatures. A resistance thermometer with $R_0 = 100\Omega$ and $\beta = 4·0 \times 10^{-3}$ K^{-1} is used to measure body temperature. If the temperature increases from 36·1 °C to 37·3 °C what is the change in resistance of the thermometer?

6.5 Skin temperature can be measured using a thermocouple because it has low thermal capacity. The potential difference, E_T, across a copper–constantan thermocouple placed at various positions along a leg is

Location	mid-thigh	knee	mid-calf	ankle	mid-foot	big toe
E_T/mV	1·28	1·07	1·15	1·07	0·99	0·79

The calibration of the thermocouple is

$T/°C$	10	15	20	25	30	35	40
E_T/mV	0·39	0·59	0·79	0·99	1·19	1·40	1·61

Using these data: (a) plot a calibration graph; (b) determine the temperature at the various locations; (c) assuming that it is your own leg, plot the temperature as a function of the distance away from the body trunk.

6.6   The respiratory rate of a subject can be measured by attaching a thermistor to the nose so that the inhaled and exhaled air flows over its tip. Describe the variation of the resistance of the thermistor during the respiratory cycle. If the time constant $\tau = 2·0$ s what is the shape of the waveform on a chart recorder? (Measure your own respiratory rate to determine T_1.)

6.7   Which of the following types of transducer can be used for isometric and isotonic measurement: (a) potentiometer displacement transducer; (b) strain gauge; (c) LVDT; (d) piezoelectric force transducer?

6.8   A DC coupled measurement system which is used to measure the force developed by a frog gastrocnemus muscle is calibrated by attaching known masses to the transducer and noting the deflection of the chart recorder display. The calibration data are given below. If the maximum deflection of the display due to the muscle twitch is 19 mm what is the force (N) developed by the muscle?

Mass/10^{-3}kg	0·5	1·0	1·5	2·0	2·5
Display/mm	8	17	23	34	39

6.9   A micropipette electrode with internal resistance $R_E = 25$ MΩ is connected to an amplifier with input resistance $R_{in} = 100$ kΩ and amplification $A_V = 100$. If the biopotential $E_B = 1·5$ mV what is the voltage at the output of the amplifier? What is the output voltage if $R_{in} = 100$ MΩ?

6.10   A $1·0$ mV$_{rms}$ signal produced by an electrode at room temperature is fed into an amplifier with bandwidth $B = 10$ kHz. Assuming that the amplifier is noise free and that when connected to the electrode the amplification is $1·0 \times 10^3$ what are the values of the signal and the noise voltages and the signal to noise ratio at the output of the amplifier? The electrode resistance is $R_E = 100$ MΩ.

6.11   What are the advantages of having an electrode with low internal resistance, R_E? Refer to Problems 6.1, 6.2, 6.9 and 6.10.

6.12   Assuming that there are no filter circuits in a measurement system describe a simple way (using a single circuit element) of eliminating high frequency noise.

6.13   An EMG signal is displayed simultaneously on a galvanometer pen recorder and an oscilloscope. Describe the difference between the signals observed on each unit.

7

Light and Lenses

Light is that part of the electromagnetic spectrum which causes the sensation of vision when it strikes the retina of the eye. A point source of light radiates electromagnetic waves in all directions (Fig. 7.1a). The speed of propagation of these waves depends upon the material through which they move. In vacuum (and air approximately) the speed of light $c = 3 \cdot 0 \times 10^8$ m s^{-1}. At a particular instant in time lines can be drawn joining the crests, or the troughs, of all the waves radiated by the source. These are called wavefronts and they form a series of concentric spheres around the source. The waves close to a point source are called spherical waves. If we consider a region far removed from the source the wavefronts are approximately parallel, as shown in Fig. 7.1b. This type of radiation is called a plane wave. Light waves from the sun are approximately plane. Waves from lamps can be made plane using mirrors and lenses.

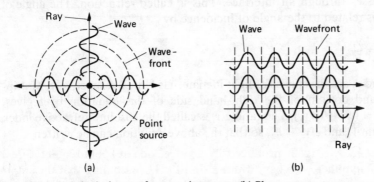

Figure 7.1 (a) Spherical waves from a point source. (b) Plane waves.

Most optical phenomena can be explained using the wave theory of light. However, the analysis is difficult and in many cases a simpler approach called ray optics can be used. A light ray is a line which indicates the direction of propagation of the waves. This line is perpendicular to the wavefronts. For spherical waves the rays radiate outwards from the source whereas for plane waves the rays are parallel (Fig. 7.1).

7.1 Reflection and Refraction

In a homogeneous isotropic material, light rays (or any electromagnetic radiation) are straight lines. If the system is not homogeneous the rays are deviated.

When the light is incident upon the interface between two media, material 1 and material 2 in Fig. 7.2, part of the radiation is reflected back into 1 and part enters into 2. If the interface has high reflectivity, such as a polished metal surface, most of the energy is reflected. For the reflected ray the angle of reflection r is equal to the angle of incidence i, $i = r$, where i and r are measured with respect to the normal direction, which is perpendicular to the interface. The incident, reflected and refracted waves and the normal are all in one plane.

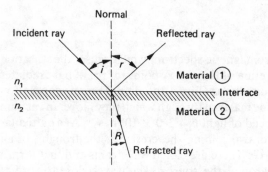

Figure 7.2   Reflection and refraction at the interface between two media.

Light travels at different speeds in different media, the speed being less in a more optically dense material. Because of this a beam of light rays is deviated when it passes through an interface. This is called refraction. The angle of refraction is related to the angle of incidence by

$$\frac{\sin i}{\sin R} = \frac{v_1}{v_2}$$

where v_1 and v_2 are the speeds of light in materials 1 and 2 respectively. Multiplying top and bottom of the right-hand side of this equation by c gives, $\sin i \sin R = (c/v_2)/(c/v_1)$. The ratio c/v_1 is called the absolute refractive index n_1 of medium 1, and $c/v_2 = n_2$, so that the above equation can be written

$$\frac{\sin i}{\sin R} = \frac{n_2}{n_1} \tag{7.1}$$

In general the refractive index n of a material is given by $n = c/v$. Values of the refractive index for some common materials are given in Table 7.1. When a light ray passes from one material to another which has a higher refractive index (e.g. air to glass) it is deviated towards the normal. The larger the value of n the larger the deviation.

Table 7.1   Refractive indices.

Material	n	Material	n
Air	1·00	Crown glass	1·52
Water	1·33	Immersion oil	1·52
Glycerine	1·47	Flint glass	1·63

Many objects are visible because they reflect or scatter incident light. If the surface is not smooth the body acts as a secondary source and radiates light in all directions. If the body is situated in a material of high refractive index, such as a fish in water, the rays are deviated away from the normal when they pass into air, and the apparent depth H', observed by the eye, is less than the actual depth H. The reason for this is shown in Fig. 7.3a, where S is the actual source. The rays from S are refracted at the interface so that they appear to come from S'.

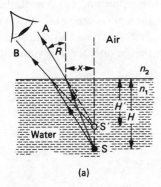 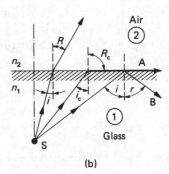

(a)   (b)

Figure 7.3 (a) The image of an underwater object. (b) Total internal reflection at the surface of a cover slip.

The light scattered by a microscope specimen, the source S in Fig. 7.3b, is refracted by the glass cover slip and only a small portion of the available light enters the microscope objective to form the final image. The angle of refraction R increases with increasing angle of incidence. A situation exists, ray A, where the angle of refraction is 90 degrees. The angle i_c at which this occurs is the critical angle of incidence. For values of $i > i_c$, ray B, reflection takes place at the interface and the light does not enter into material 2. This is called total internal reflection. At the critical condition $i = i_c$ and $R = 90$ degrees. From Equ. 7.1

$$\frac{\sin i_c}{\sin 90°} = \frac{n_2}{n_1} \tag{7.2}$$

If medium 1 is glass of refractive index 1·52 and medium 2 is air, $n_2 = 1·00$, the critical angle of incidence is given by

$\sin i_c = 1·00/1·52 = 0·659$

$i_c = 41$ degrees

7.2 Image formation by surfaces and lenses

When a parallel beam of light rays (a plane wave) is incident upon the spherical interface between two media, which have refractive indices n_1 and n_2 (with n_1 less than n_2), the rays are refracted towards the normal as shown in Fig. 7.4a. The normal passes through the centre of curvature C of the spherical surface.

The angle of refraction R is related to the angle of incidence i by Equ. 7.1. The incident rays are deviated so that they all pass through the point F_2 on the principal axis. This is called the focal point of the surface and the distance f_2 between F_2 and the pole of the surface A is called the focal length.

A curved surface is able to form an image of a source of light rays. This can be shown by constructing a ray diagram as in Fig. 7.4b. Consider an object, distance u from A, which emits light rays in all directions. Rays parallel to the principal axis are deviated towards the focal point F_2. Rays directed towards the centre of curvature C are undeviated because the angle of incidence is zero. The image I is formed where rays from the same point on the object, the tip in this case, coincide. Provided the angles of incidence are small all rays from this point will be deviated by the surface so that they pass through I and form the tip of the image. Rays from other points on the object are deviated in the same way.

The distance v between the image and A is called the image distance. Using Equ. 7.1 it can be shown that u, v and f_2 are related by

$$\frac{n_2}{v} - \frac{n_1}{u} = \frac{n_2}{f_2} \tag{7.3}$$

For a spherical surface of radius r the focal length is

$$f_2 = \frac{rn_2}{n_2 - n_1} \tag{7.4}$$

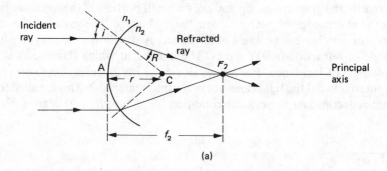

(a)

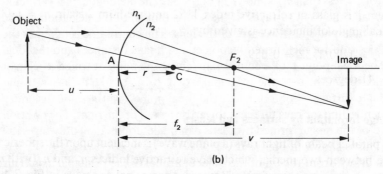

(b)

Figure 7.4 (a) Deviation of parallel rays through the focal point F_2 of a surface. (b) Image formation by a surface.

The ability of a surface to converge or diverge a beam of rays can be described by the power P. For a curved surface $P = n_2/f_2$, where the unit of P is the dioptre and f_2 is in metres.

When applying equations 7.3 and 7.4 it is necessary to employ a sign convention. A convention that is easy to use is to take the pole A as the origin of a coordinate system and measure all distances from this point. Distances measured in the same direction as the incident light are positive and distances measured against the incident light are negative. For example, if Fig. 7.4 represents an air–glass interface, then n_2 is greater than n_1 and r and f_2 are positive quantities. For light incident on a concave surface, r and f_2 are negative. With the convex interface shown u is a negative quantity and v a positive quantity.

Example 7.1

What is the power of the cornea of the eye, which is a curved surface with radius of curvature $r = 7 \cdot 8$ mm, $n_1 = 1 \cdot 000$ (air) and $n_2 = 1 \cdot 336$? If an object is placed 250 mm in front of this surface where is the image?

The power of the surface is, from Equ. 7.4

$$P = \frac{n_2}{f_2} = \frac{n_2 - n_1}{r}$$

$$P = \frac{1 \cdot 336 - 1 \cdot 000}{7 \cdot 8 \times 10^{-3}\text{m}} = 43 \text{ dioptres}$$

From Equ. 7.3, $n_2/v = n_2/f_2 + n_1/u$ so that

$$\frac{n_2}{v} = 43 \text{ m}^{-1} + \frac{1 \cdot 000}{-2 \cdot 5 \times 10^{-1} \text{ m}} = 39 \text{ m}^{-1}$$

$$v = \frac{1 \cdot 34}{39 \text{ m}^{-1}} = 34 \text{ mm}$$

A lens is a piece of transparent material with at least one curved surface. A converging lens, with refractive index n_2, immersed in a medium of refractive index n_1 (with n_1 less than n_2) is shown in Fig. 7.5a. The incident light is refracted at the front and the back surfaces. If the lens is thin all of the deviation can be assumed to take place at the centre of the lens. The centre line is taken as the reference position for measuring the object and image distances and the focal length f. The focal point F of a converging lens is the point at which a beam of rays parallel to and near to the principal axis is brought to a focus (the same as for a single surface). For a thin lens the focal points are the same distance f from the centre line for rays directed towards the left or the right. The position and nature of the image formed by a lens can be determined by drawing a ray diagram. This is shown in Fig. 7.5b.

A ray is drawn from the tip of the object parallel to the principal axis. This ray is deviated through the focal point F. A ray directed through the geometric centre of the lens is not deviated because at this point the lens acts as a parallel-

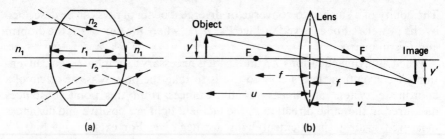

Figure 7.5 (a) Deviation of light rays at the front and back surfaces of a lens. (b) The formation of an image by a thin lens.

sided slab of material (see Problem 7.1). The image is formed where these two rays coincide. Examples of ray diagrams for a converging and a diverging lens are given in Fig. 7.6. By considering the lens as two spherical surfaces it can be shown from equations 7.3 and 7.4 that u, v and f are related by

$$\frac{1}{v} - \frac{1}{u} = \frac{1}{f} \tag{7.5}$$

where u and v are the object and image distances and

$$\frac{1}{f} = (n_2 - n_1) \left(\frac{1}{r_1} - \frac{1}{r_2} \right) \tag{7.6}$$

where r_1 and r_2 are the radii of curvature of the front and back surfaces. The linear magnification m produced by the lens, defined as the ratio of the image size to the object size y'/y (see Fig. 7.5b), is

$$m = \frac{y'}{y} = \frac{v}{u} \tag{7.7}$$

The power of a lens $P = 1/f$

When these equations are used to determine the position, nature and size of the image it is necessary to employ the sign convention described previously. Distances measured from the centre of the lens in the direction of the incident light are positive and distances measured against the incident light are negative. In Fig. 7.5b, u is a negative quantity and v is a positive quantity. For the converging lens shown in Fig. 7.5a, r_1, measured from the pole of the front surface, is positive; r_2 is negative. If the value of m given by Equ. 7.7 is positive the image is upright, as in Figs. 7.6a and c, and if it is negative the image is inverted (Fig. 7.5b). Whenever the lens equations are used to analyse an optical system an accurate ray diagram should be drawn to check the answers.

In Fig. 7.6a the rays diverge after passing through the lens. If the object is viewed by the eye, as shown, the rays appear to come from I. This is called a virtual image because the rays are not actually brought to a focus by the lens. In this case the final deviation of the rays to form the image is done by the eye. A real image is formed (such as on a screen) when the rays are actually brought together by the lens, as in Fig. 7.5b.

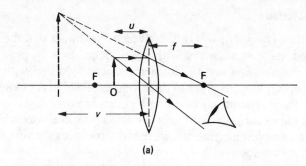

(a)

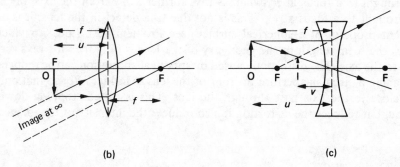

(b)   (c)

Figure 7.6 (a) and (b) Virtual images formed by a converging lens. (c) A virtual image formed by a diverging lens.

Example 7.2

An object is placed (a) 13 mm and (b) 20 mm in front of a converging lens of focal length 20 mm. What are the position and nature of the images (see Figs. 7.6a and b).

(a) $u = -13$ mm, $f = 20$ mm so that from Equ. 7.5

$$\frac{1}{v} = \frac{1}{20 \times 10^{-3} \text{ m}} + \frac{1}{-13 \times 10^{-3} \text{ m}} = 5 \cdot 0 \times 10^{1} \text{ m}^{-1} - 7 \cdot 7 \times 10^{1} \text{ m}^{-1}$$

$$1/v = -2 \cdot 7 \times 10^{1} \text{ m}^{-1}$$

$$v = -37 \text{ mm}$$

The negative sign indicates that I is on the left-hand side of the lens. The magnification $m = v/u = (-37 \text{ mm})/(-13 \text{ mm}) = 2 \cdot 8$ so the image is erect, and is virtual.

(b) $u = -20$ mm, $f = 20$ mm so that

$$\frac{1}{v} = \frac{1}{20 \times 10^{-3} \text{ m}} + \frac{1}{-20 \times 10^{-3} \text{ m}} = 0$$

$$v = \infty$$

The image is formed in infinity, as indicated by the parallel rays.

7.3   Image distortion

The lens equation $1/v - 1/u = 1/f$ is derived by assuming that the light rays are close to and make small angles with the principal axis. In practice these conditions are not satisfied and the image produced by a lens with spherical surfaces is distorted. Various types of distortion exist which are called aberrations. Some of these aberrations occur with monochromatic light (i.e. a single colour) and some occur only with light containing several wavelengths (e.g. white light). The simplest aberrations are monochromatic spherical aberration and chromatic aberration.

With spherical aberration rays close to the principal axis, ray A in Fig. 7.7a, are brought to a focus at F_A whereas rays further away from the axis, ray B, are brought to a focus at F_B. This is not due to a defect in the lens, it is an inherent property of a spherical surface (see Problem 7.3). Equ. 7.6 which relates the focal length to the geometry of the lens is only true for rays very close to the axis. The distortion caused by spherical aberration can be reduced by placing a circular aperture in front of the lens as in Fig. 7.7b, so that only rays close to the axis pass through the lens. This is called stopping down. Although it reduces the distortion it also reduces the intensity of the image.

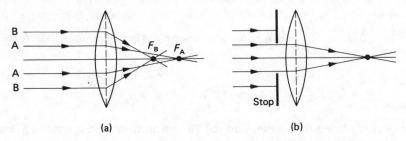

Figure 7.7   (a) Spherical aberration. (b) Stopping down to reduce distortion due to spherical aberration.

The refractive index n of a material is not a constant but varies with the wavelength (i.e. colour) of the light. Since the focal length depends on n (Equ. 7.6), the focal length and the linear magnification vary with the wavelength. When white light is incident upon a single converging lens (Fig. 7.8a) the low wavelength blue light is brought to a focus closer to the lens than the high

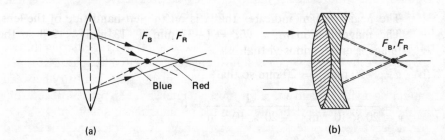

Figure 7.8   (a) Chromatic aberration. (b) Elimination of chromatic aberration using two lenses (an achromatic doublet).

wavelength red light. This is called chromatic aberration and results in an indistinct image which has coloured edges. It can be reduced by using two lenses, a converging and a diverging lens, made from glass of different refractive index as shown in Fig. 7.8b. The chromatic aberration of one lens counteracts that of the other. This combination is called an achromatic doublet. Usually compensation for chromatic aberration is only achieved for the extremes of the visible spectrum, i.e. red and blue light.

Other aberrations exist which cause distortion of the image. They can be reduced, or eliminated completely for certain colours by proper design of the lens surfaces, choice of lens material and the use of stops.

7.4   Image formation by the eye

The optical system of the eye is shown schematically in Fig. 7.9. The eyeball is approximately spherical in shape with a diameter of about 23 mm. There are three refracting surfaces, the cornea and the two surfaces of the lens. These deviate the incident light rays so that an image of an object is formed on the retina. The electrical impulses generated by the receptors in the retina are transmitted to the brain through the optic nerve and cause the sensation of vision. The normal eye is sensitive to radiation with wavelengths between 400 nm and 700 nm. The maximum sensitivity is at approximately 550 nm which is yellow-green in colour (§ 4.3).

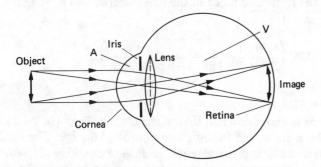

Figure 7.9   Image formation by the eye.

The cornea is a transparent layer of material 0·5 mm thick with radius of curvature 7·8 mm. The region A between the lens and the cornea is filled with a liquid called the aqueous humour which has refractive index 1·336. The average value of the refractive index of the lens material is 1·413. The vitreous humour, which fills the space V between the lens and the retina, has the same refractive index as the aqueous humour. The amount of light entering the eye is controlled by the iris which is almost opaque. The circular aperture formed by the iris is called the pupil. The diameter of the pupil can vary between 2 mm for bright light and 8 mm for low light intensity. This process of controlling the intensity of the light falling on the retina is called adaption.

In order to see an object clearly a sharp image of the object must be formed on the retina. For an optical system with fixed radii of curvature, and hence optical power, a sharp image will only be formed for one object distance. A normal relaxed eye will focus parallel rays from an object at infinity. The power of the relaxed eye is about 60 dioptres. For closer objects the power of the eye must increase in order to produce a sharp image. This is achieved by means of the ciliary muscles attached to the eye lens which increase the radius of curvature of the lens surfaces. This process is called accommodation. The range of accommodation of a normal young eye is about 10 dioptres falling to about 1 dioptre in old age. For small object distances the power of the eye is insufficient to produce a clear image on the retina. The nearest position for distinct vision is called the near point D. Values of D range from about 0·1 m for young people to about 2 m for old people. For the normal average eye D is taken as 250 mm.

Example 7.3

In its relaxed state the eye lens has radii of curvature $r_1 = 10$ mm and $r_2 = -6\cdot0$ mm (note the sign convention). If the lens is immersed in a medium of refractive index 1·336 what is the power of the lens?

The power P of the lens is $P = 1/f$. From Equ. 7.6

$$P = (n_2 - n_1)\left(\frac{1}{r_1} - \frac{1}{r_2}\right)$$

where $n_2 = 1\cdot413$ is the refractive index of the lens material and $n_1 = 1\cdot336$.

Substituting values

$$P = (1\cdot413 - 1\cdot336)\left(\frac{1}{10 \times 10^{-3}\ m} - \frac{1}{-6 \times 10^{-3}\ m}\right)$$

$$P = 21 \text{ dioptres}$$

The power of the lens is about one half that of the cornea (Example 7.1), so that the cornea is the main image forming component of the eye. The total power of the two systems is approximately $P = P_1 + P_2 = 43 + 21 = 64$ dioptres. This is close to the value given by a more exact analysis.

7.5 Interference of light

The position and nature of an image formed by a curved surface can be determined using the laws of reflection and refraction and by assuming that in a homogeneous medium light rays travel in a straight line. There are certain optical phenomena, such as interference, diffraction and polarization, that cannot be explained in this way and can only be accounted for by assuming that light is a transverse wave in which the electric and magnetic fields vary in time and space. The electric field intensity E of a light wave propagating in the z direction with speed c can be represented by

$$E = E_0 \sin\left[(2\pi/\lambda)(ct - z)\right]$$

where E_0 is the amplitude and λ the wavelength of the wave. A beam of monochromatic light consists of a bundle of waves which have the same wavelength. If in addition to having the same wavelength the phase relationship between the waves remains constant the light is said to be coherent. For a single wave the light intensity I sensed by the eye is proportional to the amplitude squared, $I \propto E_0^2$. For a beam of coherent waves I is proportional to the square of the sum of the amplitudes, account being taken of the sign.

Two coherent in-phase waves with amplitudes E_{10} and E_{20} are shown in Fig. 7.10a. At some instant, which we might take as $t = 0$, these waves can be represented by

$$E_1 = -E_{10} \sin(2\pi z/\lambda); E_2 = -E_{20} \sin(2\pi z/\lambda)$$

When these waves are added together the resulting wave has an electric field intensity $E = -(E_{10} + E_{20}) \sin(2\pi z/\lambda)$ and the light intensity $I \propto (E_{10} + E_{20})^2$. The addition of coherent waves to give a maximum in the light intensity is called constructive interference.

Some of the waves in a beam of coherent light can be shifted in time, or space, with respect to others (§ 2.4). Consider two waves, one of which has been shifted by an amount $\lambda/2$ along the z axis with respect to the other. When these two waves add the resulting light intensity $I \propto (E_{10} - E_{20})^2$. If $E_{10} = E_{20}, I = 0$ and there is darkness. The addition of waves to give a minimum or zero light intensity is called destructive interference.

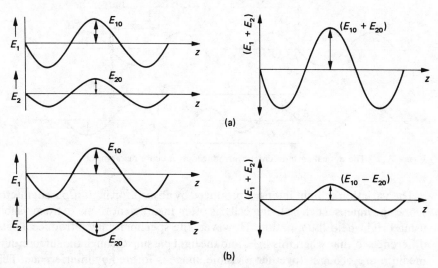

Figure 7.10   The addition of coherent light waves: (a) in-phase, $\delta = 0$; (b) out-of-phase, $\delta = \lambda/2$.

The spatial shift between the waves is called the optical path difference δ. For constructive interference the path difference must be an integral number of whole waves and for destructive interference it must be an odd number of half wavelengths:

$$\delta = m\lambda \qquad m = 0, 1, 2, 3, \ldots \quad \textit{maximum intensity}$$
$$\delta = (m + \tfrac{1}{2})\lambda \quad m = 0, 1, 2, 3, \ldots \quad \textit{minimum intensity}$$
(7.8)

Light waves from an ordinary lamp suffer from random fluctuations in phase and are coherent for only short periods of time (about 10^{-8} s). This time is much shorter than that required by the eye, or a photographic film, to detect the radiation. Coherent waves can be obtained from a single light beam by dividing it into two parts using a semi-reflecting mirror, called a beam splitter. By using a suitable arrangement of mirrors and lenses the two parts can be brought together so that they interfere. An optical path difference can be introduced during the propagation of the beams along different paths.

Two coherent waves R_1 and R_2 propagating through different media are shown in Fig. 7.11. Assume that R_1 propagates in air ($n_1 = 1 \cdot 00$) and R_2 propagates in a material with refractive index n_2 greater than n_1. The time required by R_2 to travel distance L, from B to D, is $t = L/v_2$ where v_2 is the speed of propagation in material 2. During this time R_1 will have travelled a distance $S = tc = Lc/v_2 = Ln_2$. After the two waves have reached C and D they move at the same speed and have the same wavelength. The path difference, δ, introduced by material 2 is

$$\delta = (S - L) = Ln_2 - L = L(n_2 - 1) \tag{7.9}$$

If $\delta = \lambda/2$ the two emergent beams interfere destructively when they are added together. If $E_{10} = E_{20}$ there will be zero light intensity.

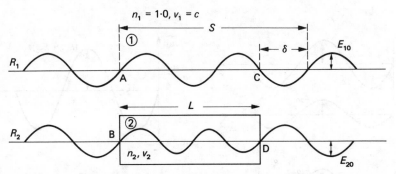

Figure 7.11   The optical path difference introduced by a dense medium.

The variation in light intensity produced by absorption in transparent microscope specimens, such as living cells, is often insufficient to make them visible using bright-field illumination. However, the specimen may introduce a path difference so that when this light and the light passing through the surrounding medium are brought together a visible image is formed by interference. This is the basis of phase contrast microscopy. (A phase shift is equivalent to a path difference.)

Example 7.4

The smallest path difference that will produce a visible image is estimated to be $\lambda/4$. What is the thickness of a specimen of refractive index $1 \cdot 4$ that will produce an image using green light with $\lambda = 546$ nm assuming that the surrounding medium is air?

From Equ. 7.9 the path difference $\delta = (n - 1)L = \lambda/4$ so

$$L = \frac{\lambda}{4(n - 1)} = \frac{546 \times 10^{-9} \text{ m}}{4(1 \cdot 4 - 1)}$$

$$L = 0 \cdot 34 \, \mu\text{m}$$

Path differences can be introduced into a beam of light by reflection from the top and bottom surfaces of a thin transparent film. If the reflected waves are brought together using a lens, or by the eye, they interfere. A thin film of refractive index n_2 on a substrate of refractive index n_3 greater than n_2 is shown in Fig. 7.12. Some of the waves in the incident beam are reflected at the air–film surface and some are refracted. The refracted waves, R_2, travel a longer distance than the reflected waves, R_1. There will be constructive or destructive interference between R_1 and R_2 according to the optical path difference δ (Equ. 7.8). When the angle of incidence is small, i.e. near normal incidence, the conditions are

$$\begin{array}{lll} 2n_2d = m\lambda & m = 0, 1, 2, 3 \ldots \textit{constructive} \\ 2n_2d = (m + \tfrac{1}{2}) \lambda & m = 0, 1, 2, 3 \ldots \textit{destructive} \end{array} \qquad (7.10)$$

where d is the thickness of the film. These conditions take into account the fact that there is a phase change of 180 degrees in the waves when they are reflected from a more dense medium. In the arrangement of Fig. 7.12 there are 180 degree phase changes at A and B, which cancel out. If the refractive index of the substrate is less than that of the film there is no phase change at B and the conditions for constructive and destructive interference are the reverse of those given in Equ. 7.10. If the thin film is illuminated with white light the conditions for constructive and destructive interference may be satisfied for some wavelengths but not others and the film appears coloured.

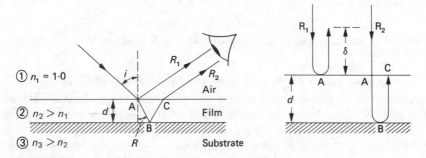

Figure 7.12 Reflection from the top and bottom surfaces of a thin film.

Example 7.5

A $0 \cdot 31 \, \mu$m thick film of water, $n = 1 \cdot 33$, in air is illuminated with white light at normal incidence. What colour does it appear in reflected light?

Maximum light intensity will be obtained at the wavelength for which there is constructive interference. Since in this case there is no substrate $n_3 = n_1$

and the conditions are the reverse of those given in Equ. 7.10. Thus there will be maximum intensity at the wavelengths given by

$$\lambda = \frac{2dn}{m + \frac{1}{2}} = \frac{2 \times 310 \text{ nm} \times 1 \cdot 33}{m + \frac{1}{2}}$$

Substituting values for $m(= 0, 1, 2, 3 \ldots)$ shows that there will be constructive interference at wavelengths 1650 nm, 550 nm, 330 nm, etc. Since the only maximum intensity in the visible region is for light with wavelength 550 nm the film appears to be green in reflected light.

7.6   Diffraction

When a beam of light passes through a narrow aperture some of the waves are deviated into the region of the geometric shadow. This is called diffraction and is due to the interference of waves from a large number of coherent sources.

If a parallel beam of light is incident upon a slit, which has width less than the wavelength of the light, the slit acts as a secondary point source and radiates waves in all directions (Fig. 7.13a). When a larger slit is illuminated (Fig. 7.13b), each point on the aperture acts as a new source of coherent waves. If waves originating from these sources, travelling in a particular direction θ, are brought together by a lens they interfere constructively or destructively according to their path difference. The path difference depends upon the angle θ.

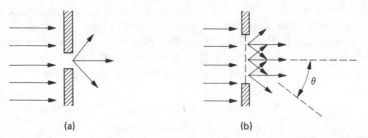

(a)   (b)

Figure 7.13   A parallel beam of light incident on a slit. (a) Slit width less than λ. (b) Slit width about the same as λ.

Considering Fig. 7.14a, waves from the aperture travelling in the $\theta = 0$ direction are in phase and they add constructively to produce brightness. The wave R_1 (Fig. 7.14b), originating from point A on the aperture travelling in direction θ_1, has a path difference λ with respect to R_3 originating from point C. Since R_1 from A has a path difference $\lambda/2$ with respect to wave R_2 from B, R_1 and R_2 interfere destructively when they are brought together. For any wave from the region AB of the aperture there is a wave from the region BC which has path difference $\lambda/2$. The intensity of the image formed by the interference of waves travelling in this direction is zero. The angle θ_1 for this first diffraction zero is

$$\sin \theta_1 = \lambda/d \tag{7.11}$$

The conditions for constructive and destructive interference are satisfied alternately as the diffraction angle θ increases so that the image of a narrow

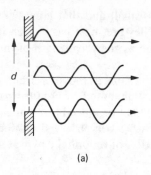
(a)

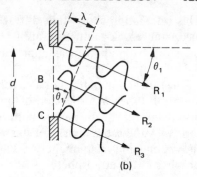
(b)

Figure 7.14 (a) Constructive interference of waves radiated at $\theta = 0$; brightness. (b) Destructive interference of waves radiated at θ_1; darkness.

slit consists of a bright line, the central maximum ($\theta = 0$), with alternating dark and bright bands on either side. The intensities of the secondary maxima are much less than the central maximum.

The diffraction pattern produced by a circular aperture consists of a bright central disc, called Airy's disc, surrounded by alternate dark and bright diffraction rings. This is shown in Fig. 7.15. For an aperture with diameter d the angle θ_1 for the first diffraction zero is given by

$$\sin \theta_1 = 1 \cdot 22 \, \lambda/d \qquad (7.12)$$

which differs from Equ. 7.11 by the numerical factor $1 \cdot 22$.

For a circular aperture with diameter 40 mm illuminated with light of wavelength 580 nm the angle for the first diffraction zero is given by

$$\sin \theta_1 = \frac{1 \cdot 22 \times 580 \times 10^{-9} \, \text{m}}{4 \cdot 0 \times 10^{-2} \, \text{m}} = 1 \cdot 8 \times 10^{-5}$$

Since θ_1 is very small $\sin \theta_1 \approx \theta_1$ radians. If the light is focussed onto a screen using a lens of focal length $f = 1 \cdot 0$ m and assuming that $L \approx f$, the radius r_1 of the first dark ring is

$$r_1 = L\theta_1 = 1 \cdot 0 \, \text{m} \times 1 \cdot 8 \times 10^{-5} = 1 \cdot 8 \times 10^{-5} \, \text{m}$$

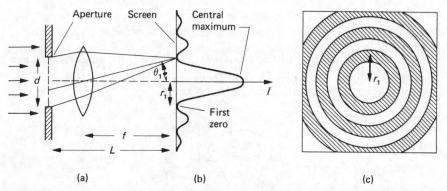

(a)   (b)   (c)

Figure 7.15 Diffraction at a circular aperture. (a) The geometric arrangement. (b) The intensity distribution across the screen. (c) The image on the screen.

This calculation shows that diffraction effects are small and that great care has to be employed if they are to be observed. They are most easily seen if the size of the aperture is approximately the same as the wavelength of the light.

The diffraction pattern of a circular obstacle is the same as that of a circular aperture. When a suspension of small particles is illuminated by light from a distant source the diffraction pattern is the same as that produced by an aperture with the same dimensions as the particles. The haloes observed around the moon on a misty night are due to the diffraction of moonlight by water droplets in the atmosphere.

7.7   Resolution

Since a lens has finite dimensions it acts like a circular aperture and the image of a point source is a diffraction pattern. When several sources are close together, such as in a specimen observed under a microscope, the diffraction patterns associated with each source overlap and the image is indistinct. The ability of a lens system to show clear detail of the object is described by its resolving power. The resolving power is the minimum distance between two points in an object that can be seen clearly in the image. The Rayleigh resolution criterion states that the limit of resolution is when the first diffraction zero of one source coincides with the central maximum of the other (Fig. 7.16).

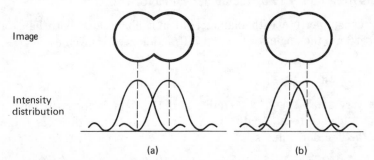

Figure 7.16   The appearance of the image and the intensity distribution for two point sources; (a) just resolved; (b) not resolved.

Example 7.6

The diameter of the pupil of the eye is 4·0 mm. What is the smallest distance between two point objects at the near point such that they are just resolved? Assume that the wavelength of the light is 500 nm.

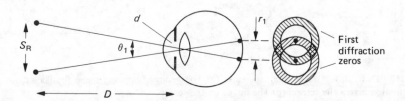

If the first diffraction zero of one pattern falls on the central maximum of the other the angular separation of the sources is θ_1 where θ_1 is given by Equ. 7.12. (This is the Rayleigh resolution criterion.) Thus

$$\sin \theta_1 = 1\cdot22\lambda/d$$

$$\sin \theta_1 = \frac{1\cdot22 \times 5\cdot0 \times 10^{-7} \text{ m}}{4\cdot0 \times 10^{-3} \text{ m}} = 1\cdot5 \times 10^{-4}$$

Since θ_1 is small $\sin \theta_1$ approximately equals $\tan \theta_1$ which approximately equals θ_1 radians. The limit of resolution, the distance S_R, is

$$S_R = D\theta_1 = 2\cdot5 \times 10^{-1} \text{ m} \times 1\cdot5 \times 10^{-4} = 38 \ \mu m$$

This value of S_R is the resolving power of the optical system of the eye. In order to see the images as distinct the network of receptors in the retina must be fine enough to record these separate images. That is, the resolving power of the eye is governed by two factors; the resolving power of the optical system and the packing of the receptors in the retina. Maximum resolving power, which is usually taken as approximately 100 μm at the near point, is obtained with a pupil diameter of about 4 mm. The visual acuity is less at larger diameters because of aberrations, and at smaller diameters because of diffraction effects.

7.8 Polarized light

Interference and diffraction occur with any type of wave but the phenomenon of polarization can only be observed with transverse waves. Light can be polarized, indicating that it is a transverse wave whereas sound cannot, indicating that it is a longitudinal wave.

A light wave is usually represented diagrammatically by a sine curve. The line joining the axis to the peak of the curve is called the wavevector. In Fig. 7.10 the wavevectors E_{10} and E_{20} are oriented in the y direction. This is a special case and in general the vector can take any direction provided it is perpendicular to the direction of propagation z. If we view a beam of light along the z axis the electric vectors will be randomly oriented in the yx plane (Fig. 7.17a)

All of the electric vectors can be removed from a beam of light except those in one particular direction (Fig. 7.17b). This is called linearly polarized light and

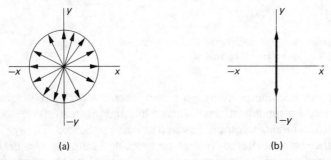

(a)   (b)

Figure 7.17 (a) The electric vectors in ordinary light. (b) The electric vector in linearly polarized light.

the direction of the electric vector E is the direction of polarization. Any vector E at an angle θ to the y direction can be resolved into components $E_x = E \sin \theta$ and $E_y = E \cos \theta$ along the x and y directions. The x and y components of all waves in a beam add together so that ordinary light can be represented by two mutually perpendicular, linearly polarized vectors of equal magnitude. Partially polarized light is a mixture of ordinary and polarized light. It can be represented by two perpendicular vectors of unequal size.

The simplest way of producing polarized light is to pass ordinary light through a material called Polaroid. This is a thin sheet of plastic, such as polyvinyl alcohol, in which the large molecules are oriented in a particular direction. When the electric vector of the incident light is parallel to the direction of molecular orientation it is absorbed and when it is perpendicular to the molecular orientation it is transmitted. This preferential absorption of polarized light is called dichroism. The direction of transmission within the Polaroid sheet is the polarization direction.

When polarized light is observed through a Polaroid the intensity of the transmitted light depends upon the angle θ between the direction of polarization of the light and the polarization direction of the Polaroid. Consider the arrangement shown in Fig. 7.18. The first polarizing element is called the polarizer and the second is called the analyser. The electric vector E_p of the polarized light produced by the polarizer can be resolved into two perpendicular components, $E_\parallel = E_p \cos \theta$, parallel to the polarization direction OA of the analyser and $E_\perp = E_p \sin \theta$ perpendicular to OA. $E_\parallel$ is transmitted by the polarizer and $E_\perp$ is absorbed. The intensity I of the transmitted light is proportional to $E_\parallel^2$ and can be written as

$$I = I_0 \cos^2 \theta$$

where I_0 is the intensity when $\theta = 0$. When $\theta = 90$ degrees the transmitted light intensity is zero. This is called the extinction position and the analyser is crossed with the polarizer.

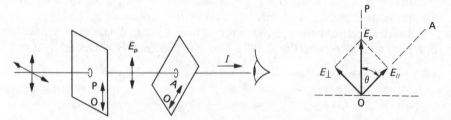

Figure 7.18 The production and analysis of linearly polarized light. OP and OA are the polarization directions of the polarizer and analyser.

A light beam can become partially polarized when it is reflected from a surface or scattered by small particles. When white light is incident on small particles the scattered light is coloured as well as being polarized. The visible radiation from the sun is scattered by dust particles and air molecules in the earth's atmosphere (Fig. 7.19). The sky overhead, viewed along the y direction appears blue rather than black because low wavelength blue waves in the sunlight are scattered at right angles to the direction of propagation of the incident beam. The longer wavelength red waves are unaffected. At sunset, when the sun is viewed directly along the x-axis it appears red because the blue waves have been scattered out of the forwards direction. The state of polarization of the scattered light can be determined by viewing the blue sky through a Polaroid. When the Polaroid is rotated the transmitted intensity passes through a maximum and a minimum. The direction of polarization of the scattered light is perpendicular to its direction of propagation and the direction of the sunlight. This property of scattered sunlight can be used as a navigational aid (instead of a compass). Honey bees are able to orient themselves in flight using polarization sensing elements in their eyes which detect the state of polarization of the light from the sky.

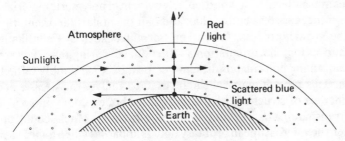

Figure 7.19   Scattering of light by the earth's atmosphere.

The state of polarization of light is changed when it passes through certain types of solid or liquid. The simplest of these effects is the rotation of the direction of polarization. This is known as optical activity. The optical rotation α can be to the right (positive) or to the left (negative) looking back along the direction of propagation. For optically active solutions

$$\alpha = [\alpha]_\lambda^T Lc$$

where $[\alpha]_\lambda^T$(rad kg^{-1} m^2) is the specific rotation for a particular wavelength λ and temperature T, L(m) is the optical path length through the solution and c(kg m^{-3}) is the concentration.

Optical activity is due to asymmetry in the atomic arrangement. In some materials, such as proteins, carbohydrates and steroids the asymmetry is associated with the molecules. With other types of material, such as crystalline solids, it is associated with the packing of the molecules and disappears when the solid is melted or put into solution.

The optical rotation produced by a solution can be measured using a polarimeter (Fig. 7.20). The sample column is initially filled with the solvent and the analyser crossed with the polarizer. The sample is introduced and the analyser again set to extinction. The angular rotation of the analyser is the optical rotation introduced by the sample. This type of measurement is used for the quantitative analysis of solutions (composition and concentration) and for the study of molecular structure.

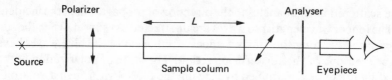

Figure 7.20   A polarimeter.

With some materials the refractive index depends upon the direction of polarization of the incident light. These materials are called doubly refracting, or birefringent. When a doubly refracting material is placed between crossed polars light is transmitted by the analyser. If white light is used the image may be coloured. Double refraction, like optical activity, is due to the anisotropic packing of atoms and molecules. It is particularly evident in polymeric materials.

In natural substances such as cotton, flax and hair the molecular orientation is a result of the growth process. In synthetic products such as Cellophane, viscose rayon and nylon the orientation is introduced during manufacture. Some transparent amorphous materials, e.g. Perspex, become doubly refracting under stress. This type of material can be used to analyse the stresses and strains within models of structural elements such as bones. Many biological materials are transparent but exhibit double refraction. Much detail not observable under the microscope in ordinary light is apparent when a specimen is viewed through a crossed polarizer–analyser system.

Problems

7.1   Light rays are incident on a parallel-sided glass slab of refractive index $n_2 = 1.52$ at an angle $i = 60$ degrees. If the slab is 10.0 mm thick and is immersed in air ($n_1 = 1.00$) calculate using Equ. 7.1 the angle at which the rays emerge from the slab. Draw a ray diagram showing the relationship between the incident and transmitted rays.

7.2   Remembering that for a plane wave the wavefronts are perpendicular to the rays, draw a diagram showing that plane waves are produced when a point source is placed at the focal point of a converging lens.

7.3   Draw an arc of a circle 50 mm radius on millimetre graph paper. Assume that this is the interface between air ($n_1 = 1.00$) and glass of refractive index $n_2 = 1.52$. Construct the principal axis and draw two rays parallel to the axis 20 mm and 50 mm away from the axis. Using Equ. 7.1 draw the refracted rays and determine the positions where they cross the axis. They should both cross at the focal point but they do not because of spherical aberration. Compare these values with that given by Equ. 7.4. The normal to the surface always passes through the centre of the circular arc.

7.4   Comment on the position, nature and size of an image formed by a converging lens when the object is: (a) outside the focal length; (b) at the focal point; and (c) inside the focal length. Illustrate your answers using ray diagrams.

7.5 A relaxed eye can be represented by a single refracting surface of power 60·0 dioptres, with $n_1 = 1·000$ and $n_2 = 1·336$. What is: (a) the focal length (see Fig. 7.4); and (b) the image distance for an object at infinity? Does the image fall on the retina? Illustrate your answer with a ray diagram.

7.6 At age 40 years the accomodating power of the normal eye is about 4·0 dioptres and the near point 250 mm. What is: (a) the total optical power of the accomodated eye (refer to Problem 7.5); and (b) the image distance for an object at the near point? Illustrate your answer with a ray diagram. How does this compare with the answer to Problem 7.5? Comment.

7.7 Show that when a specimen of refractive index n_2 and length L is immersed in a medium of refractive index n_1 the path difference between the light passing through the surrounding medium and the waves scattered by the specimen is $\delta = (n_2 - n_1)L$, see Equ. 7.9. If $L = 5·0 \ \mu m$ and $n_1 = 1·4$ what is the smallest difference in refractive index between specimen and immersion medium such that the specimen is visible in green light, $\lambda = 546 \ nm$, using the phase contrast technique.

7.8 The coloured appearance of the inside of an oyster shell is not due to pigmentation. Explain.

7.9 Light from a laser with wavelength 632 nm is directed towards glass plate upon which a fine powder has been evenly distributed. The diameter D_1 of the first dark ring of the diffraction pattern formed on a screen, distance L from the plate, is measured as a function of L. Assuming that the angle θ_1 for the first diffraction zero is given by Equ. 7.12 calculate from the following data the mean diameter of the powder particles.

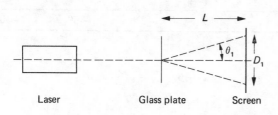

L/m	0·40	0·57	0·79	0·95	1·18	1·45
$D_1/10^{-2} m$	2·0	2·7	3·9	4·6	6·1	7·4

7.10 The distance between the centres of two small red warning lights on a machine is 10·0 mm. Assuming that these lights are point sources and that the resolution is limited by diffraction only, determine the maximum distance an operator can be from the machine so that he can still see these lights clearly.

7.11 When ordinary light is reflected from a transparent material such that the angle between the reflected and refracted beams is 90 degrees the reflected beam is linearly polarized. The direction of polarization is parallel to the interface. Some sunglasses are made from Polaroid sheet. How should the polarization direction be oriented to minimize the intensity of light reflected from a wet road?

7.12　The outer segments of the rods and cones in the eyes of vertebrates consist of a stack of membranous discs. The visual pigment, which is a long chain molecule, is randomly oriented in the plane of these discs as shown below. When viewed in a direction perpendicular to the axis using polarized light the stack exhibits dichroism. What is the direction of polarization for strong absorption? Is the stack dichroic along the axis?

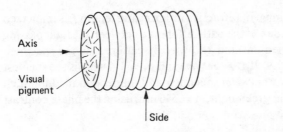

7.13　A polarimeter, with column length 0·10 m, is used to measure the optical rotation for sodium D light produced by solutions of collagen with different concentration c (g/100 ml of solution). Using the following data determine: (a) the specific rotation of collagen (in SI units); and (b) the concentration of an unknown solution of collagen, 0·15 m long, which produces an optical rotation of -17 degrees.

$c/\text{g}(100\text{ ml})^{-1}$	2	4	6	8	10
α/degrees	-8	-15	-24	-30	-40

8

Microscopes

A knowledge of the internal structure of an organism is necessary to understand the function of the organism and the way in which it lives and grows. This structure can be determined by visual observation. The apparent size of an object seen by the eye is determined by the size of the image formed on the retina and depends upon the angle subtended at the front of the eye on the principal axis by light rays from the object. This is shown in Fig. 8.2a. The angle α can be increased and a larger image produced if the object is brought closer to the eye. Since the unaided eye cannot see clearly objects closer than the near point D the largest useful image is obtained when the object is at this point. If a converging lens is placed in front of the eye the object can be brought closer than D and still remain in focus on the retina. A lens used in this way is called a simple magnifier. If the magnification produced by a single lens is insufficient several lenses can be used. This is called a microscope. The purpose of magnifiers and microscopes is to increase the angle α at which the light enters the eye so as to produce a clearly defined magnified image of the object. The approximate dimensions of some biological organisms and the type of microscope used for their examination are given in Fig. 8.1.

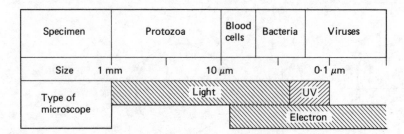

Figure 8.1   Some biological organisms and the microscopes used for their examination.

8.1   The simple magnifier

The simple magnifier consists of a single lens placed between the object and the eye as shown in Figs 8.2b and c. The lens forms a virtual image I′ of the object O somewhere between the near point D and infinity. The two extreme situations are shown in the diagram. Since the size of the retinal image is proportional to the angle at which the light rays enter the eye, the effective magnification γ of the lens can be defined as the ratio of the angle (α' or α'') subtended at the eye using the magnifier to the angle (α) subtended at the unaided eye if the same object were at the near point. This is also called the angular magnification and should not be confused with the linear magnification m (Equ. 7.7). When I' is at infinity $\tan \alpha' = y/f$ and $\tan \alpha = y/D$. Since α and α' are small $\tan \alpha' \approx \alpha'$ and $\tan \alpha \approx \alpha$ so that

$$\gamma = \frac{\alpha'}{\alpha} = \frac{yD}{fy} = \frac{D}{f} \tag{8.1}$$

If the virtual image I' is at the near point and the lens is close to the eye $\gamma = \alpha''/\alpha = (y/u)(D/y) = D/u$. From Equ. 7.5, $-1/u = (1/f - 1/v)$ where $v = -D$ (note the sign convention) so that $1/u = -(D + f)/Df$. Ignoring the minus sign, which indicates that the object is on the left-hand side of the lens, we have

$$\gamma = \frac{D}{u} = 1 + \frac{D}{f} \tag{8.2}$$

When the image is at the near point the linear magnification m is the same as the angular magnification.

Lieuwink's simple microscope consisted of a single, almost spherical, bead lens with radius approximately 1·0 mm. Using Equ. 7.6 with $r_1 = 1·0$ mm, $r_2 = -1·0$ mm and $n = 1·5$ the focal length is calculated to be about 1·0 mm. From Equ. 8.1 the effective magnification of this lens is approximately 250 ×. The symbol × is used to indicate magnification. Usually aberrations set an upper practical limit of about 20 × for a single lens.

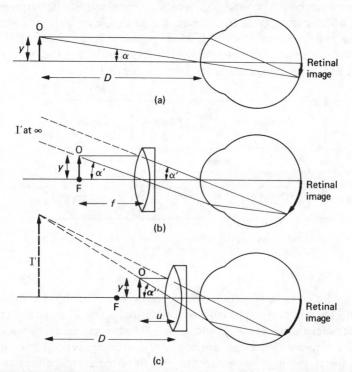

Figure 8.2 Image formation using a simple magnifier. (a) The unaided eye. (b) Virtual image I' at infinity. (c) Virtual image I' at the near point.

8.2 The compound microscope

Lens aberrations can be reduced and a larger useful image obtained by employing a combination of lenses. The optical components of a compound microscope are shown in Fig. 8.3a. The image forming part consists of two multi-lens systems, the objective which has short focal length and the eyepiece (sometimes

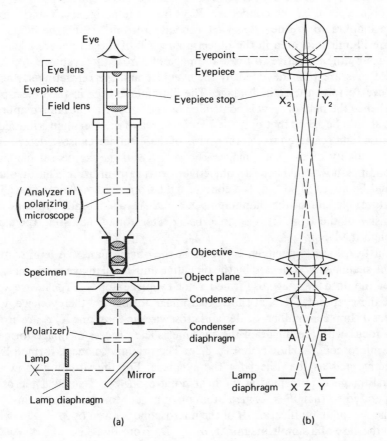

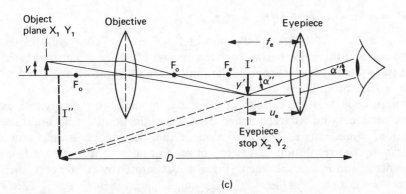

Figure 8.3   (a) The optical components of a compound microscope. (b) The path of rays from the lamp diaphragm. (c) Image formation in a microscope.

called the ocular) which has a longer focal length. They are about 160 mm apart. An adequate understanding of the operation of a microscope can be obtained by treating the lens systems as thin lenses.

The path of the rays from a light source in a microscope set up with Kohler illumination is shown in Fig. 8.3b. Rays from the axial point Z in the lamp diaphragm XY fill the condenser diaphragm AB and are brought to a focus in the object plane by the condenser. Rays from the periphery X and Y of the lamp diaphragm are deviated by the condenser and define the limits X_1Y_1 of the illuminated area in the object plane. All of these rays are brought to a focus by the objective to form an image of XY at the eyepiece stop X_2Y_2. These rays are then deviated by the eyepiece and the refracting surfaces of the eye to form the retinal image. The illuminated area in the object plane, and hence the field of view is controlled by the size XY of the lamp diaphragm. This should be set so that only the required area of the specimen is illuminated because light scattered from other parts of the specimen causes a deterioration in the quality of the final image. The size AB of the condenser diaphragm determines how much of the objective aperture is utilized. This diaphragm should be adjusted so that the cone of light from the condenser just fills the aperture. Neither of the diaphragms XY and AB are intended to control the intensity of the light. This is done using filters or by adjusting the current through the lamp.

If a translucent specimen is placed in the object plane the light scattered by the specimen is collected by the objective and a real image of the specimen, called the primary image, is formed at the eyepiece stop X_2Y_2. The eye, which is at the eyepoint, looks at the virtual image formed by the eyepiece, which acts as a simple magnifier. Scales and crosswires in the plane X_2Y_2 are brought to a focus with the image of the specimen. The process of image formation in a microscope is shown in Fig. 8.3c. The primary image I' formed by the objective acts as the object for the eyepiece. For the situation shown, the virtual image I'' is formed at the near point of the eye. Usually a microscope is operated so that I'' is formed at infinity in order to avoid eye strain.

The overall magnification M of the microscope is given by $M = \alpha''/\alpha$ where, as in the case of a simple magnifier, $\alpha = y/D$. From Fig. 8.3c, $\tan \alpha'' = y'/u_e = y'(D + f_e)/Df_e$. The size of the primary image $y' = m_0 y$ where m_0 is the linear magnification of the objective. Since $\alpha'' \approx \tan \alpha''$ the overall magnification is

$$M = \alpha''/\alpha = m_0(1 + D/f_e) = m_0 \gamma_e \qquad (8.3)$$

where γ_e is the effective magnification of the eyepiece. The values of the magnification of the objective (m_0) and the eyepiece (γ_e) are usually indicated on the lens system. For the eyepiece the value given is the effective magnification $\gamma_e = D/f$, for an image at infinity.

A microscope must be capable of resolving fine detail in the specimen as well as producing an enlarged image. High magnification is of no use if the resolution is inadequate. The resolution of the microscope is determined primarily by the resolution of the objective. Consider light scattered from two points P_1 and P_2 in a specimen, (Fig. 8.4a), separated by a distance S_R such that they are just resolved (see § 7.7). The angle θ_1 is given by $\sin \theta_1 = 1 \cdot 22 \lambda/d$ where d is the diameter of the objective aperture. It can be shown that if the medium between the specimen and the objective has refractive index n the resolving power is $S_R = 1.22 \lambda/2n \sin u$ where u is the semi-angle of the cone of rays collected by the objective. The product $n \sin u$ is called the numerical aperture (NA) so that the resolving power is

$$S_R = 0.61 \lambda/\text{NA} \qquad (8.4)$$

The larger the value of NA the smaller the value of S_R and the better the resolution. For an objective of NA = 0.6 the smallest distance between two points in a specimen that will be resolved in green light, $\lambda = 546$ nm, is $S_R = 0.61 \times 546 \times 10^{-9}$ m/0.6 = 0.56 μm.

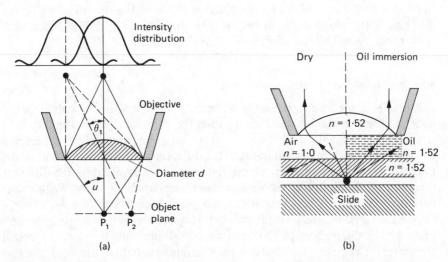

Figure 8.4 (a) Resolution of the images formed by an objective. (b) The collection of light by a dry and an immersion objective.

Better resolution (i.e. a smaller value of S_R) can be obtained by decreasing λ and by increasing NA. The range of wavelengths available in the visible spectrum is limited and does not allow much variation in the resolution. If ultraviolet light is used and the image made visible photographically the resolving power can be increased by about a factor of two. The maximum value that $\sin u$ can take is 1.0. For dry objectives where the medium between the specimen and the front objective lens is air (Fig. 8.4b), the maximum theoretical value of NA is unity. In practice the largest value that can be achieved is about 0.95. If an oil, with n greater than 1.0, is introduced into the space between the specimen and the objective a significant increase in NA, and resolution, is obtained. Most high magnification objectives are of the immersion type and have NA up to about 1.5. If the oil used with the objective has the same refractive index as the glass cover slip there is no refraction at the cover slip–oil interface (Fig. 8.4b). Consequently an immersion objective is able to collect more light from the specimen than a dry objective and produces a brighter image. The intensity of illumination of the image is approximately proportional to $(\text{NA})^2$.

A microscope is able to form a sharp image of objects which are small distances above or below the object plane so that the object plane is not a surface but is a layer. The thickness of this layer, which is called the depth of focus, is proportional to $1/(NA)^2$. The conditions for high resolution, high illumination and large depth of focus are not compatible.

If the magnification of a microscope is too large the eye will see an unresolved image of the specimen and the image will contain spurious detail. The useful magnification M^* of a microscope is that magnification which produces a separation S_E in the final image, I″, of two points in the object plane separated by a distance S_M where S_E is the resolving power of the eye and S_M the resolving power of the microscope. That is, $M^* S_M = S_E$ or $M^* = S_E/S_M$. In practice $S_E \approx 0 \cdot 1$ mm. For a high power objective of NA = $1 \cdot 5$ the resolving power S_M (Equ. 8.4) is about $0 \cdot 22$ μm so that M^* is approximately 500 ×. Magnifications about twice this are often used to avoid eye strain.

8.3   Specialist light microscopes

Features in a specimen are visible because of variations in the colour or the light intensity in the image. This is called the contrast. In the ordinary light microscope contrast is obtained because different parts of the specimen absorb light by different amounts. This differential absorption modifies the intensity (i.e. the amplitude) of the light waves transmitted by the different regions and the image is visible because of amplitude contrast. With transparent specimens, such as unstained living cells, there is no differential absorption so that a visible image is not formed. However, with this type of specimen optical path differences (§ 7.5) are introduced into waves that pass through the different regions. When these path differences (which are equivalent to phase differences, §2.4) are converted to amplitude differences a visible image is obtained. This is done in the phase contrast and the interference microscopes.

The principle of operation of a phase contrast microscope is shown in Fig. 8.5. The object O is illuminated by a parallel beam of light which has wavelength λ. If the dimensions of features in the object are about the same as λ some of the light is diffracted and is brought to a focus in the image plane. The diffracted light is shown as the broken line. The undiffracted light, the full line, is brought to a focus in the plane P. It then diverges to form a uniformly illuminated background in the image plane. Since the object and surrounding medium, or different parts of the object, have different refractive index (n_2 and n_1 respectively) there is an optical path difference $\delta = L(n_2 - n_1)$ between the diffracted and undiffracted waves, where L is the thickness of the specimen. Usually δ is approximately $\lambda/4$. The amplitude and phase relationships in the diffracted and undiffracted waves are shown in Fig. 8.5b. In order to obtain a visible image it is necessary to introduce a further $\lambda/4$ phase shift into the waves, so that they are exactly out of phase, and to reduce the amplitude (intensity) of the undiffracted waves. This is done by inserting a glass phase ring in the plane P. The intensity of the undiffracted waves is reduced by coating the central portion of the phase ring with a thin layer of metal. The amplitude and phase relationships between the waves at the image plane is shown in Fig. 8.5c. When the beams interfere to form an image the intensity in the interference pattern depends upon the path difference δ introduced by the object. Differences in the refractive index n_2, or thickness L, of the features in the specimen give rise to differences in the intensity (i.e. amplitude contrast). Image contrast is greatest for regions of rapidly changing refractive index so that fine detail is overemphasized. This difficulty is overcome in the interference microscope.

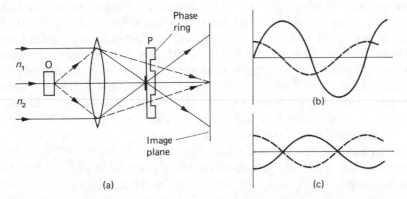

Figure 8.5 (a) Ray paths in the phase contrast microscope. (b) The amplitudes and phase relationship in front of the phase ring. (c) The amplitudes and phase relationship at the image plane. The full lines are the undiffracted rays and the broken lines are the diffracted rays.

In the interference microscope, shown in Fig. 8.6a, two coherent beams of light are formed by dividing the incident beam into two parts using a beam splitter (a semi-transparent mirror). One part is directed through the specimen O, which has refractive index n_2, and the other travels through air, $n_1 = 1 \cdot 0$. The two beams, which have path difference $\delta = L(n_2 - 1)$, are recombined and the image formed by interference. This technique does not rely upon diffraction to separate the beams and is able to reveal much finer detail than the phase contrast method.

Structural detail in transparent objects can also be made visible using polarized light (the polarizing microscope) and by using dark-ground illumination. With dark-ground illumination (Fig. 8.6b), the specimen is illuminated by

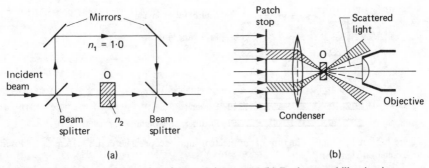

Figure 8.6 (a) The ray path in an interference microscope. (b) Dark-ground illumination.

a highly divergent hollow cone of light. This is obtained by placing a patch stop in front of the condenser. None of the direct light enters the objective. The image is formed by light scattered from small structural inhomogeneities in the specimen. This is shown as the broken line.

8.4   The electron microscope

The ability of a light microscope to reveal the detailed structure of a specimen is limited by its resolving power. The maximum resolving power of a light microscope is about 0·2 μm and the useful magnification about 1000 ×. Electrons have a dual nature (§ 3.2). An electron is usually considered to be a small particle having mass $M_e = 9\cdot11 \times 10^{-31}$ kg and a negative charge of $-1\cdot60 \times 10^{-19}$C. In some situations however, a beam of electrons behaves like a bundle of waves and can be reflected, refracted and diffracted. These properties of an electron beam are utilized in the electron microscope. Since the wavelength of an electron beam is much less than that of visible light a much higher resolution can be obtained (Equ. 8.4).

A simplified diagram of an electron microscope is shown in Fig. 8.7. A divergent beam of electrons is obtained from an electron gun (§6.5). This beam is brought to a focus in the vicinity of the specimen by the condenser

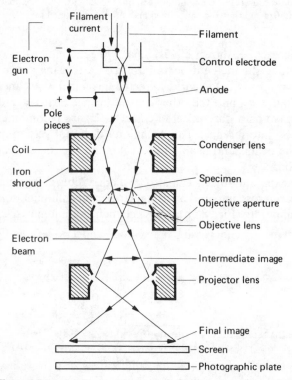

Figure 8.7   A schematic diagram of an electron microscope. The entire system is enclosed in an evacuated chamber.

lens. The lenses in most electron microscopes consist of a coil of wire enclosed in a soft-iron shroud. When a current is passed through the coil a magnetic field is produced between the pole-pieces. This field deviates the electron beam in much the same way as a glass lens deviates a beam of visible light. When the electron beam passes through the specimen some of the electrons, shown by the broken lines, are scattered out of the beam and are blocked by the objective aperture. The electrons that pass straight through are deviated by the objective lens to form an intermediate image in front of the projector lens. This then forms a real image on a flat fluorescent screen. When the electrons strike the screen light waves are emitted and a visible image is seen. The intensity of different parts of the image, i.e. the contrast, depends on how many electrons are scattered out of the incident beam by the different features in the specimen. A region that is transparent to electrons is bright and a region that has high scattering power is dark. The screen can be removed so that the electrons fall onto a photographic plate and a permanent record of the image obtained.

The wavelength λ of the electron waves is related to the accelerating potential V by $\lambda = h/\sqrt{2M_e eV}$ where $h(= 6\cdot63 \times 10^{-34}$ J s$)$ is Planck's constant. For an accelerating potential of 100 kV the wavelength is about 4×10^{-3} nm. Electromagnetic lenses suffer from the same aberrations as optical lenses. In the electron microscope these are reduced to a minimum by using apertures which confine the electron beam to a small region close to the principal axis. Because of this the angular aperture of the objective (u in Fig. 8.4a) is small. The numerical aperture NA of an electron microscope is usually about 1×10^{-2}. For a 100 kV microscope the resolving power, $S_R = 0\cdot61$ $\lambda/$NA, is approximately $0\cdot3$ nm, which is 10^3 times better than that of a light microscope. In order to make full use of this resolution the magnification $M^* = S_E/S_M$ should be about $1 \times 10^{-4}/3 \times 10^{-10} = 330\,000$.

In the ordinary light microscope image contrast is obtained mainly through the process of absorption whereas in the electron microscope it is due to scattering. Most of the electron scattering is caused by collisions with the electron clouds surrounding the nuclei of the atoms in the specimen. Heavy atoms, with large atomic number, and hence large electron cloud, have a higher scattering power than light atoms. Biological specimens are usually made up of atoms with low atomic number, e.g. carbon and hydrogen, (see Table 3.2) so that the image contrast is low. The contrast can be increased and detail made visible by treating the specimen with metal salts which preferentially stain different regions.

Problems

8.1 An object 10 mm high is situated at the near point of the eye ($D = 250$ mm) so that an image is formed on the retina. If the diameter of the eyeball is 25 mm what is the size of the retinal image? What will be the size of the retinal image if a magnifier of focal length 100 mm is placed in front of the eye the virtual image I′ being formed at infinity?

8.2 Show that when a simple magnifier forms a virtual image at the near point the linear magnification $m = v/u = (1 + D/f)$, which is the same as the angular magnification.

8.3 Two common defects of the eye are short sight and long sight. In short sight parallel rays from a source at infinity are brought to a focus by the relaxed eye in front of the retina so that only objects close to the eye can be seen clearly. In long sight parallel rays are brought to a focus behind the retina so that objects can only be clearly seen if they are a long distance away from the eye. Draw ray diagrams for these two situations. What type of spectacle lens, converging or diverging, should be used to correct these defects?

8.4 Show by means of a ray diagram how the size of the illuminated area in the object plane of a compound microscope is controlled by the size of the lamp aperture.

8.5 When a compound microscope is set up to form a virtual image I″ at infinity the image I′ formed by the objective is at the front focal point, F_e, of the eyepiece (see Fig. 8.3c). Draw a ray diagram showing how the image I″ is formed.

8.6 Find the minimum distances between two point objects such that their images are just resolved by an objective for: (a) green light, $\lambda = 500$ nm; (b) violet light, $\lambda = 420$ nm; (c) ultraviolet light, $\lambda = 253$ nm; and (d) electron waves, $\lambda = 1.0 \times 10^{-2}$ nm. Take typical values of NA given in the text.

8.7 Calculate for the following objectives the smallest distance S_R between two features in a specimen such that their images are just resolved: (a) $m_0 = 4\times$, NA $= 0.13$; (b) $m_0 = 40\times$, NA $= 0.70$; (c) $m_0 = 100\times$, NA $= 1.3$ (oil immersion). These objectives are used with an eyepiece marked 10 ×, the image I″ being formed at the near point of the eye. Assuming that $\lambda = 500$ nm determine the distance between these features in the image I″. (Refer to Problem 8.5.) Are these images resolved by the eye?

8.8 Two images of the same object are produced at the same magnification using objectives with NA of 0.7 and 1.4. Which image is the brightest? What is the ratio of the intensity of illumination of the images?

8.9 Discuss briefly, and comment on, the conditions necessary for high resolution, high image illumination and large depth of focus.

8.10 Explain why it is not possible to see a homogeneous transparent object if it is immersed in a medium which has the same refractive index.

8.11 The difference in the refractive index of two regions in a tissue cell is 0.10. If the thickness L of the specimen is 2.0 μm what is the optical path difference δ for green light of wavelength 546 nm travelling through these regions? Express your answer in wavelengths.

9

Thermal and Surface Properties

Living organisms are complex arrangements of different types of material. The important properties of these materials are specific to their function in the organism. For example structural materials such as bone are strong and tough, transport materials such as blood are easily moved from one location to another, body fluids such as water are good solvents for the chemicals necessary for the maintainance of life. These materials have properties not related to their primary function which also influence the way in which the organism behaves. Because of this it is necessary to know what the properties are and how they can be measured.

Materials can exist in three forms: solid, liquid and gas. These are called the three states of matter. Each state has characteristic properties. In solids the molecules are bound together by strong attractive forces so that they are fixed in a particular location. The molecules can be arranged in a regular way with respect to each other over long distances (long-range order). Solids are able to sustain a shear stress (see § 10.1). In gases the attractive intermolecular forces are small so that the molecules are free to move within the containing vessel. Gases are completely disordered and cannot sustain a shear stress. The liquid state is intermediate between the solid and the gaseous states. Liquids exhibit short-range order (over a few molecular diameters), as in solids, but cannot exhibit long-range order. The molecules in a liquid are relatively free to move from one location to another and a liquid cannot sustain a shear stress.

The differences in the properties of the different states are due to two antagonistic factors; the strength of the intermolecular bonds, which hold the molecules together, and the magnitude of the thermal energy, which tends to disrupt these bonds. When the thermal energy is small the molecules are strongly bound together by the intermolecular forces and the material is a solid. When the thermal energy is large the molecules are not bound and the material is a gas. In this chapter we first discuss the relationship between thermal energy and temperature and then consider the thermal and surface properties of materials. The properties of structural and transport materials are described in the following chapter.

9.1 Temperature and molecular motion

Many of the properties of a gas are accounted for by a simple theory which assumes that the gas molecules are infinitely small non-interacting elastic spheres. (Neither of these assumptions is true in practice.) Consider a vessel containing gas molecules which move in all directions. When these molecules collide with the walls of the vessel (Fig. 9.1a), they experience an acceleration because their direction of movement has changed. Consequently the walls of the vessel exert a force on the molecules and from Newton's third law the molecules must exert a force on the walls. The pressure P of the gas is the force per unit area due to these collisions. It can be shown from the simple kinetic theory of gases that for a vessel containing N molecules per unit volume

$$P = \tfrac{2}{3}N(M_0 v^2/2)_{av} \tag{9.1}$$

where $(M_0 v^2/2)_{av}$ is the average value for the kinetic energy of a single molecule, which has mass M_0 and speed v.

For a gas at low pressure, the pressure P is related to the volume V of the gas by the equation

$$PV = nRT \tag{9.2}$$

where n is the number of kmol of gas, R is the gas constant ($8 \cdot 31 \times 10^3$ J kmol^{-1} K^{-1}) and T (K) is the absolute temperature. The units of P are N m^{-2} (or Pa). By combining equations 9.1 and 9.2 it can be shown (Problem 9.1) that

$$(M_0 v^2/2)_{av} = \tfrac{3}{2}kT \tag{9.3}$$

where $k = R/N_A$ is Boltzmann's constant ($1 \cdot 38 \times 10^{-23}$ J K^{-1}) and N_A is Avogadro's constant ($6 \cdot 02 \times 10^{26}$ kmol^{-1}).

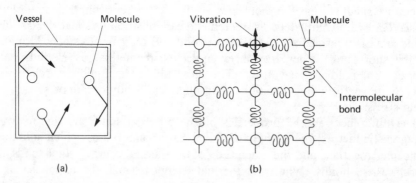

Figure 9.1 (a) The motion of gas molecules inside a container. (b) The bonding between molecules in a crystalline solid. The bonds are drawn as springs to indicate the nature of the vibrational motion.

Equ. 9.3 enables us to define quantitatively what is meant by temperature. (The value of temperature indicated by a thermometer is arbitrary because it depends upon the scale that is used. The sensations of hot and cold, which are associated with high and low temperatures, are subjective and cannot be quantified.) The kinetic energy of the gas molecules, which is called the internal heat energy of the gas, is directly proportional to the absolute temperature of the gas. It is called absolute temperature because at zero degrees absolute

(0 K) the internal energy is zero. A temperature of less than zero degrees absolute cannot exist. On the Celsius temperature scale zero absolute is $-273 \cdot 15$ °C. Celsius temperatures are related to absolute temperatures by the equation

$$T(°C) = T(K) + 273 \cdot 15 \ (K)$$

The same concept applies to gases, liquids and solids. In a solid there are strong intermolecular forces so that the molecules are not free to move around, as they are in gases, but remain in fixed positions (Fig. 9. 1b). However, they are able to vibrate about these fixed positions. The internal heat energy of a solid is the kinetic energy of vibration. In general we can say that heat energy, which is proportional to the absolute temperature, is equivalent to molecular kinetic energy.

9.2 Specific heat capacity

When heat energy is put into a gas, liquid or solid, say by applying a flame to it, the internal energy and temperature of the material increase. The heat energy input ΔQ (J) is related to the change in temperature $\Delta T(K)$ by

$$\Delta Q = MC\Delta T \tag{9.4}$$

where M is the mass of the material and $C(J \ kg^{-1} \ K^{-1})$ is called the specific heat capacity.

The value of C depends on the constraints placed upon the material. If the volume remains constant (as for a gas in a vessel of fixed size) the quantity involved is C_V, the specific heat capacity at constant volume. With solids the process usually occurs at constant external (atmospheric) pressure. When a solid is heated its volume increases so that a quantity of work is done by the material on the surroundings. Part of the heat energy input is used to increase the heat energy of the material and part is used to perform work. Consequently in order to produce the same temperature change a larger quantity of heat is required in a constant pressure process than in a constant volume process. The specific heat capacity at constant pressure C_P is larger than C_V. For solids and liquids the values quoted usually refer to constant pressure. Values for some common substances are given in Table 9.1.

Table 9.1 Specific heat capacities at constant pressure.

Substance	$C_P/10^3 \ J \ kg^{-1} \ K^{-1}$	Substance	$C_P/10^3 \ J \ kg^{-1} \ K^{-1}$
Water	$4 \cdot 19$	Wood	$1 \cdot 76$
Body tissue	$3 \cdot 35$	Air	0.99
Ice	$2 \cdot 11$	Copper	$0 \cdot 39$

If a quantity of heat ΔQ is introduced into a system in a time Δt the rate of temperature change $\Delta T/\Delta t$ is given by

$$\frac{\Delta Q}{\Delta t} = MC\frac{\Delta T}{\Delta t} \tag{9.5}$$

where $\Delta Q/\Delta t$ is in watts (W = J s^{-1}) and the units of $\Delta T/\Delta t$ are K s^{-1}.

Example 9.1

For a man under basal conditions the rate of metabolic heat production ($\Delta Q/\Delta t$) for unit body surface area is about $44 \cdot 4$ W m^{-2}. What would be the rate of temperature rise, ($\Delta T/\Delta t$), of a 70 kg man with surface area $1 \cdot 80$ m^2 if there were no loss of heat energy to the surroundings?

For a man with surface area $1 \cdot 80$ m^2 the rate of heat production is

$$\Delta Q/\Delta t = 1 \cdot 80 \text{ m}^2 \times 44 \cdot 4 \text{ W m}^{-2}$$

$$\Delta Q/\Delta t = 80 \text{ W}$$

If the man is completely insulated from the surroundings all of the metabolic heat is used to raise the temperature of the body tissue. For $M = 70$ kg and $C = 3 \cdot 35 \times 10^3$ J kg^{-1} K^{-1}.

$$\frac{\Delta T}{\Delta t} = \frac{\Delta Q/\Delta t}{MC}$$

$$\frac{\Delta T}{\Delta t} = \frac{80 \text{ J s}^{-1}}{70 \text{ kg} \times 3 \cdot 35 \times 10^3 \text{ J kg}^{-1} \text{ K}^{-1}}$$

$$\Delta T/\Delta t = 3 \cdot 4 \times 10^{-4} \text{ K s}^{-1} = 1 \cdot 2 \text{ °C/hour}$$

The man would survive for only a few hours with this rate of temperature increase.

There are several ways in which heat energy is transferred from body tissue to the surrounding environment. The various heat inputs and heat outputs from a living system are shown schematically in Fig. 9.2. At equilibrium, i.e. when the temperature of the system is constant, there must be an energy balance, the total heat energy input being equal to the total heat energy output. We will consider the ways in which the system can lose heat.

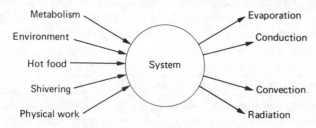

Figure 9.2   The heat inputs and outputs from a living system.

9.3   Fusion and vaporization

When heat energy is put into a solid the vibrational energy of the molecules increases. When the energy of the molecules is larger than the intermolecular bond energy the molecules break their bonds and are free to move (Fig. 9.3a). The solid has turned into a liquid. This process is called melting, the reverse is called fusion, and the temperature T_M at which it occurs is called the melting point. Since the liquid has a higher internal energy than the solid, heat energy must be given to the solid to cause it to melt. The quantity of heat energy involved in the phase transition is proportional to the mass M of material, the proportionality constant being called the latent heat of fusion, L_F,

$$Q = ML_F \tag{9.6}$$

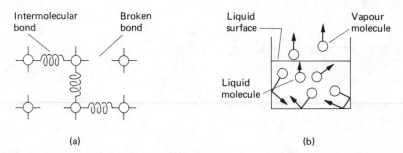

Figure 9.3 (a) Melting of a solid due to the breaking of intermolecular bonds. (b) Evaporation from the surface of a liquid.

In the liquid state the molecules are free to move around within the space defined by the vessel and the surface of the liquid. The heat energy is not evenly distributed among the individual molecules, some molecules have higher energy than others. These high energy molecules are able to break through the surface of the liquid and form a vapour (Fig. 9.3b). This process is called vaporization or evaporation. Since the molecules left behind in the liquid have, on average, less heat energy than those that escape the temperature of the liquid decreases during evaporation. The temperature of a body in contact with the liquid will also decrease. The quantity of heat involved in this phase transition is given by Equ. 9.6 with L_F replaced by the latent heat of vaporization L_V. M is the mass of the liquid that has evaporated. Values for the latent heats of fusion and vaporization and the melting and boiling temperatures of some common substances are given in Table 9.2. The high values of L_F for ethanol and water are due to the relatively strong (hydrogen) intermolecular bonds in these materials. The latent heats of vaporization given in Table 9.2 are the values at the boiling point. L_V increases with decreasing temperature.

Table 9.2 Latent heats of fusion and vaporization.

Substance	T_M/°C	L_F/10^5 J kg^{-1}	T_B/°C	L_V/10^5 J kg^{-1}
Oxygen	−219	0·14	−183	2·14
Ethanol	−114	1·04	78	8·55
Mercury	−39	0·12	357	2·72
Water	0	3·33	100	22·6
Copper	1629	1·34	1733	50·7

Example 9.2

The human body loses about 0·030 kg of water per hour by evaporation (from the sweat glands and from the lungs). What is the rate of energy loss?

Converting all quantities to SI units the rate of energy loss due to evaporation is, from Equ. 9.6,

$$\frac{\Delta Q}{\Delta t} = L_V \frac{\Delta M}{\Delta t} \tag{9.7}$$

$$\frac{\Delta Q}{\Delta t} = \frac{2 \cdot 26 \times 10^6 \text{ J kg}^{-1} \times 0 \cdot 030 \text{ kg h}^{-1}}{3 \cdot 6 \times 10^3 \text{ s h}^{-1}}$$

$$\Delta Q / \Delta t = 19 \text{ W}$$

which is about one quarter of the metabolic heat production under resting conditions.

9.4   Heat conduction, convection and radiation

Heat energy can move from one location to another in three ways, by conduction, convection and radiation.

Consider a slab of material with thickness d and surface area A (Fig. 9.4). When one face of the slab is heated the vibrational energy of the molecules at this face increases so they become hot. These hot molecules collide with adjacent cold molecules and transfer some of their heat energy. These cold molecules become hot. Heat energy is transferred through the slab by intermolecular collisions. This process is called heat conduction and occurs in solids, liquids and gases. The quantity of heat energy ΔQ that flows through the slab in time Δt is proportional to the difference $(T_1 - T_2)$ in the temperatures of the two faces. If the thickness d of the slab is much smaller than the transverse, y and z, dimensions the rate of heat flow is given by Equ. 9.8 in Table 9.4, where $K(\text{W m}^{-1} \text{ K}^{-1})$ is the thermal conductivity of the material. The thermal conductivities of some common materials are given in Table 9.3.

The transfer of heat by thermal conduction can only take place when the materials involved have high thermal conductivity, such as inside the human body. If an animal is surrounded by still air, which is a good thermal insulator,

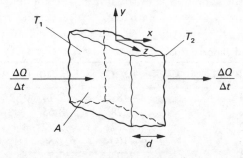

Figure 9.4   Conduction of heat through a slab of material with thickness d.

Table 9.3   Thermal conductivity.

Substance	K/W m^{-1} K^{-1}	Substance	K/W m^{-1} K^{-1}
Copper	384	Wood	0·126
Water	0·59	Air	0·025
Animal fat	0·21	Eiderdown	0·019

heat loss to the environment by conduction can only take place where the animal is in contact with a solid material, e.g. through its feet. Consequently heat losses by conduction are usually small. However, if the animal is immersed in a good thermal conductor, such as water, large conductive heat losses can occur.

Table 9.4   The heat transfer equations.

Heat transfer mechanism	Rate of heat transfer $\Delta Q/\Delta t$ (W)	
Conduction	$KA(T_1 - T_2)/d$	(9.8)
Convection	$HA(T_S - T_E)$	(9.9)
Radiation	$e\sigma A(T_S^4 - T_E^4)$	(9.10)

When a material is heated its volume usually increases because of the increased amplitude of the molecular vibrations. This is called thermal expansion. Since the mass of material involved remains the same the density of the material decreases. In liquids and gases where the molecules are free to move the hot, less dense, material rises upwards through the cold, more dense, material so that heat energy is moved from one place to another. This is called convection and is shown in Fig. 9.5a. The rate of heat transfer by convection from a surface with area A is given by Equ. 9.9, Table 9.4; H(W m^{-2} K^{-1}) is the convective heat transfer coefficient, T_S is the temperature of the surface and T_E is the temperature of the surrounding environment. The value of H depends upon the shape of the surface and the movement of the fluid in the environment. In still air H is about 4 W m^{-2} K^{-1} for humans.

Heat energy can also be transferred from one place to another in the form of electromagnetic radiation and consequently does not require a medium for its transmission. Heat energy obtained from the sun, which is transmitted through free space (approximately a vacuum) is obtained in this way. The radiation is caused by molecular vibrations and electronic transitions. At low temperatures the wavelength of the radiation is in the infrared region and is not visible. At higher temperatures the wavelength is in the red end of the visible spectrum and the material is red in colour. At very high temperatures the colour is white.

The rate of radiant heat transfer between a body, with area A and surface temperature T_S, and an enclosure with wall temperature T_E is given by Equ. 9.10. σ is Stefan's constant which has a value of $5 \cdot 67 \times 10^{-8}$ W m^{-2} K^{-4}. The emissivity e of the surface has a value between 0 and 1 according to the radiating body. For human skin e is about unity. Good emitters are also good absorbers of radiation, a perfect emitter ($e = 1$) absorbs all of the incident radiation. For non-perfect radiators e depends upon the temperature

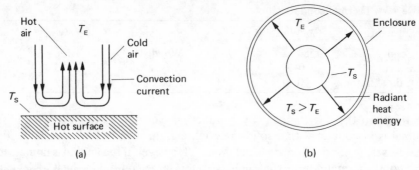

Figure 9.5 Transfer of heat by: (a) convection; (b) radiation.

and the reflectance $\rho = 1 - e$ depends upon the wavelength of the incident radiation. An effective way of maintaining body temperature in a low temperature environment is to wrap the body in metallic foil. The foil is a good reflector of radiation so that body heat is not able to escape to the environment.

Example 9.3

What is the rate of radiant energy loss of a person with surface area $1 \cdot 80$ m² and skin temperature 32°C situated in an enclosure with wall temperature 20°C?

The emissivity of human skin $e = 1 \cdot 0$, $T_S = 305$ K and $T_E = 293$ K so that from Equ. 9.10

$$\Delta Q/\Delta t = 1 \cdot 0 \times 5 \cdot 67 \times 10^{-8} \text{ W m}^{-2} \text{ K}^{-4} \times (8 \cdot 65 - 7 \cdot 37) \times 10^9 \text{ K}^4 \times$$

$$1 \cdot 80 \text{ m}^2$$

$$\Delta Q/\Delta t = 1 \cdot 3 \times 10^2 \text{ W}$$

9.5   Surface energy and surface tension

The molecules inside a liquid or a piece of solid material are attracted equally in all directions by neighbouring molecules, as shown in Fig. 9.1b, so that the net force experienced by these molecules is zero. Molecules in the surface of a solid or a liquid have no neighbours on one side so that these surface molecules experience a force (due to the unbalanced intermolecular bonds) directed towards the interior of the material. When a molecule is taken from the interior to the surface work has to be done to overcome this force and the molecule gains potential energy. The potential energy E_s (J m⁻²) per unit area of surface is called the surface energy of the material. This may be regarded as the energy required to create unit area of new surface. A surface exhibits properties which are not characteristic of the bulk material. When water flows from a narrow tube it forms small spherical drops. This is because any system will tend towards the state of lowest energy and a sphere is the shape which has the smallest surface area (and hence surface energy) for a given volume. A thin rod can float on the surface of a liquid even if it has a higher density than the liquid. It is as if the liquid has a surface skin which supports the weight of the rod. When the surface of the liquid is modified, by adding a few drops of detergent, the rod will sink.

Consider a liquid surface supported on a frame which has a rod AB that can be moved (Fig. 9.6a). Suppose the rod moves to A'B' when a constant force F is applied to it. If the temperature of the surface does not change the work, $W = F\Delta x$, done by the force is equal to the increase in the surface energy $E_s A$, where $A = L\Delta x$ is the increase in the surface area. Thus $F\Delta x = E_s L\Delta x$ or

$$E_s = F/L = \gamma \tag{9.11}$$

The surface energy, E_s (J m^{-2}), of a material is equal to the force per unit length $\gamma = F/L$ (N m^{-1}) that exists in the surface. γ is called the surface tension. Both solids and liquids possess surface energy so that they both have a surface tension. The surface of a material should not be considered as an elastic membrane. The force required to stretch a membrane increases in proportion to the deformation (§ 10.1) whereas for a surface the force is constant, independent of the deformation.

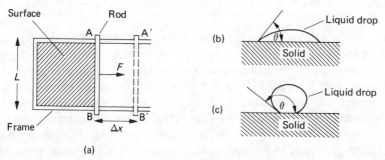

Figure 9.6   (a) A liquid surface supported by a frame. (b) A drop of liquid with contact angle θ less than 90 degrees. (c) A drop with θ greater than 90 degrees.

9.6   Contact angle and wettability

When a small quantity of liquid is placed carefully onto a solid surface it will spread across the surface until an equilibrium shape is obtained, as shown in Figs 9.6b and c. The actual shape is that which gives the lowest energy for the liquid – solid system. The angle θ between the liquid and the solid surfaces at the periphery of the drop is called the contact angle (θ is measured through the liquid). The periphery of the drop is shown in more detail in Fig. 9.7a.

The three interfaces each have their own surface energy: liquid–vapour (γ_{LV}), solid–vapour (γ_{SV}) and solid–liquid (γ_{SL}). At equilibrium the sum of the surface tension forces for unit length of interface at the solid–liquid–vapour junction is zero. γ_{LV} can be resolved into components $\gamma_{LV} \cos \theta$ and $\gamma_{LV} \sin \theta$ parallel and normal to the solid surface. Equating parallel components gives

$$\gamma_{SV} = \gamma_{SL} + \gamma_{LV} \cos \theta \tag{9.12}$$

The normal component $\gamma_{LV} \sin \theta$ is balanced by an equal and opposite adhesive

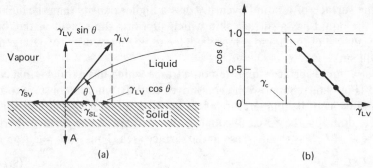

Figure 9.7 (a) The equilibrium of forces at the periphery of a drop. (b) The variation of $\cos\theta$ with γ_{LV}.

force A due to solid surface. If the solid surface is flexible a distortion in the surface can be seen around the periphery of the drop due to this normal force. The contact angles for some water – solid combinations are given in Table 9.5.

Table 9.5 The contact angles for water and solid surfaces.

Solid	Clean glass	Paraffin wax	Insect cuticle
Contact angle/degree	0	105	100 – 110

When the contact angle is zero the liquid will spread over the surface to form a thin even layer. This is called wetting. For this to occur the forces to the left in Fig. 9.7a must be larger than the forces to the right, i.e. γ_{SV} greater than γ_{SL} + γ_{LV}. Materials can be divided into two categories, high surface energy and low surface energy. Liquids and soft solids, such as organic polymers and waxes have low surface energy, less than $0 \cdot 1$ J m^{-2}. Hard materials, such as metals and glass, have high surface energy, between $0 \cdot 5$ and 5 J m^{-2}. Most liquids wet high energy solids but with low energy solids the behaviour depends upon the specific solid – liquid combination.

An empirical parameter, called the critical surface tension, which gives an indication of the surface energy of a solid can be determined by measuring the contact angle θ for a series of liquids with different values of γ_{LV}. When $\cos\theta$ is plotted as a function of γ_{LV}, as shown in Fig. 9.7b the line can be extrapolated to the point where $\cos\theta = 1 \cdot 0$, i.e. $\theta = 0$. The value of γ_{LV} at which this occurs is the critical surface tension, γ_c, of the solid surface. γ_c is the surface energy of a liquid that will just wet the surface. A liquid will spread over the surface provided γ_{LV} is less than γ_c. Some values of γ_{LV} of liquids and γ_c of low energy solids are given in Table 9.6.

Table 9.6 Surface tension of some liquids and the critical surface tension of some solids.

Liquid	$\gamma_{LV}/10^{-3}$ N m^{-1}	Solid	$\gamma_c/10^{-3}$N m^{-1}
Ethanol	22	PTFE	22
Soap solution	25	Paraffin wax	27
Olive oil	32	Polyethylene	31
Water	73	Polyvinylchloride	40

The solid – liquid interaction can be modified by altering the solid surface or the liquid. The leaves of plants are low energy surfaces so that water does not spread but remains in the form of drops. If a wetting agent, such as a detergent, is added to the water it will wet the surface. The cleansing action of a liquid depends upon its ability to wet a surface. Insect cuticle is a low energy surface which is not wet by water so that some insects can float on water supported by surface tension forces. Porous fabrics can be made waterproof by applying a thin layer of low surface energy material to the surface of the fabric.

9.7 Capillarity

When a capillary tube, with inside radius R, is held vertically with one end below the surface of a liquid, which has contact angle θ less than 90 degrees, the liquid rises up the tube and eventually reaches an equilibrium position (Fig. 9.8). This is known as capillary rise or capillarity. It can be explained in terms of the surface tension forces at the solid–liquid junction. The fixed capillary tube exerts a force γ_{LV} on the liquid. This force can be resolved into components $\gamma_{LV} \cos \theta$ in the upwards direction and $\gamma_{LV} \sin \theta$ normal to the surface of the tube. Since the liquid is free to move it will rise up the tube until the downwards force due to the weight of the column of liquid in the tube (which is shaded) is equal to the upwards force due to the surface tension. The downwards force is Mg where M is the mass of the liquid in the column. This is, M = density × volume = $\rho\pi R^2 H$. The upwards force is $\gamma_{LV} \cos \theta$ times the circumference of the circle which has radius R. At equilibrium $\gamma_{LV} 2\pi R \cos \theta = \rho\pi R^2 Hg$ so that the capillary rise H is

$$H = 2\gamma_{LV} \cos \theta / R\rho g \tag{9.13}$$

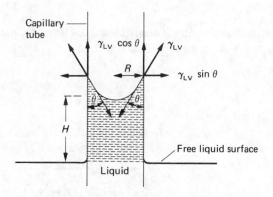

Figure 9.8   The capillary rise of a liquid in a narrow tube.

Example 9.4

What is the capillary rise of water in a tube of radius 20 μm assuming that the contact angle is zero?

For water $\gamma_{LV} = 73 \times 10^{-3}\,\text{N m}^{-1}$ and $\rho = 1\cdot0 \times 10^{3}\text{kg m}^{-3}$, so from Equ. 9.13

$$H = \frac{2 \times 73 \times 10^{-3}\,\text{kg m s}^{-2}\,\text{m}^{-1} \times 1\cdot0}{20 \times 10^{-6}\,\text{m} \times 1\cdot0 \times 10^{3}\,\text{kg m}^{-3} \times 9\cdot8\,\text{m s}^{-2}}$$

$$H = 0\cdot74\,\text{m}$$

If the contact angle between the liquid and the solid is greater than 90 degrees (e.g. mercury – glass) the liquid in the tube is depressed below the level of the free liquid surface.

9.8   Measurement of surface tension

The surface tension of a liquid can be measured by determining the capillary rise H of the liquid in a tube of known internal radius R. If the contact angle is zero, which is often the case, $\gamma_{LV} = R\rho gH/2$. The main disadvantages of this method are that the capillary must be absolutely clean and if the contact angle is not zero it must be measured.

A method that is widely used for routine measurement is to determine the maximum force required to pull a circular wire frame, radius R, through the surface of the liquid. The frame, which is usually made from a platinum alloy, is initially held below the surface of the liquid. The vessel containing the liquid is slowly moved downwards and the maximum force on the ring, just before it breaks away from the liquid surface, is measured using a force transducer. The experimental arrangement is shown in Fig. 9.9a. Since platinum is a high energy surface the contact angle is usually zero. The downwards force on the ring due to surface tension forces (Fig. 9.9b) is $2\gamma_{LV}\cos\theta$ for unit length of ring. Since the circumference of the ring is $2\pi R$ the maximum force $F = 4\gamma_{LV}\pi R \cos\theta$. For zero contact angle

$$\gamma_{LV} = F/4\pi R \tag{9.14}$$

The advantage of this method is that the ring is easily cleaned. However, the forces involved are small so that a high sensitivity force transducer (§ 6.3) must be used.

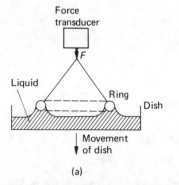

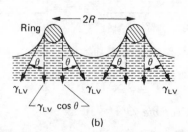

(a)   (b)

Figure 9.9   (a) Measurement of surface tension using the ring method (b) The surface tension forces on the ring.

An interesting application of surface tension forces is found in the bug Aphelocheirus. This insect breathes gaseous air but lives beneath the surface of water. The surface of the insect is covered with a down of fine stiff hairs the tips of which are bent parallel to the cuticle surface as shown in Fig. 9.10a. Surface tension forces support the pressure of the water surrounding the insect so that a layer of air is trapped between the hair tips and the cuticle (Fig. 9.10b).

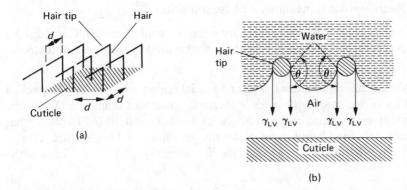

Figure 9.10 (a) The hairs on Aphelocheirus. (b) The surface tension forces on the hair tips, θ is approximately 110 degrees.

At the solid–liquid junction the hair tips exert an upwards force on the water and the water exerts a downwards force on the hairs. As the pressure is increased the line of contact between the liquid and solid moves around the circumference of the hair tip. The maximum pressure that the surface forces can sustain is when the surface tension force γ_{LV} is perpendicular to the cuticle surface as shown in the diagram. At equilibrium the upwards force due to the hairs equals the downwards force due to the water pressure. For a single hair tip with length d there are two solid–liquid junctions so that the upwards force is $2\gamma_{LV}d$. If there are N hairs per unit surface area the maximum force per unit area, i.e. the maximum water pressure $P(\text{N m}^{-2})$, that the surface tension forces can support is $2N\gamma_{LV}d$. Assuming that the length d of a hair tip is equal to the average distance between the hairs, $d = 1/\sqrt{N}$ so that the maximum water pressure is

$$P = 2N\gamma_{LV}/\sqrt{N} = 2\gamma_{LV}\sqrt{N}$$

It is estimated that $N = 2 \cdot 5 \times 10^{12}$ m^{-2} so

$$P = 2 \times 73 \times 10^{-3} \text{ N m}^{-1} \times \sqrt{2 \cdot 5 \times 10^{12} \text{ m}^{-2}}$$

$$P = 2 \cdot 3 \times 10^5 \text{ N m}^{-2}$$

The surface tension forces are able to support a maximum water pressure about 2·3 times atmospheric pressure (which is $1 \cdot 0 \times 10^5$ N m^{-2}).

Problems

9.1   Show, using equations 9.1 and 9.2, that for an 'ideal' gas $(M_0 v^2/2)_{av} = (3/2)kT$.

9.2   Write an equation that expresses the heat energy balance in a living organism in terms of the various heat energy inputs $(\Delta Q/\Delta t)_{in}$ and the heat energy outputs $(\Delta Q/\Delta t)_{out}$.

9.3   Describe what is meant by a phase transition.

9.4   Explain what happens to mercury when it is cooled to $-50\,°C$ and discuss why a mercury-in-glass thermometer cannot be used to measure low temperatures.

9.5   A dead animal of mass $1\cdot0$ kg at an intial temperature of 30 °C is packed in $1\cdot0$ kg of ice at $0\cdot0$ °C inside a thermally insulated container. When the system comes to equilibrium $0\cdot30$ kg of ice has melted. (a) Explain what is meant by equilibrium and state the equilibrium temperature of this system. (b) What is the average specific heat capacity of the body tissue of the animal?

9.6   Consider the system described in Problem 9.5. Show that if all of the ice is melted $MC(T_i - T_f) = mL_F + mc(T_f - 273)$ where M and C are the mass and the specific heat capacity of the animal, T_i is the initial temperature of the animal and T_f is the final temperature of the system, m and c are the mass and the specific heat capacity of water and L_F is the latent heat of fusion water. What is the final temperature of the system if $M = 4\cdot0$ kg?

9.7   A moving swimmer is immersed in water at 7 °C. If the thickness of the fatty tissue is 20 mm what is the rate of heat energy loss due to thermal conduction through the tissue assuming that the body core temperature is 37 °C?

9.8   What is the rate of radiant energy loss of a person with surface temperature 32 °C, surface area $1\cdot8$ m² and weight $6\cdot9 \times 10^2$ N situated in an enclosure with wall temperature 25 °C? What would be the rate of temperature decrease if there were no metabolic production of heat energy?

9.9   Consider a nude person with surface area $1\cdot8$ m² standing on a marble platform in still air. The rate of evaporation $\Delta M/\Delta t = 3\cdot5 \times 10^{-2}$ kg/h, the area of feet in contact with the slab is $4\cdot0 \times 10^{-2}$ m², the thermal conductivity, thickness and temperature difference across the tissue of the soles of the feet are $0\cdot30$ W m⁻¹K⁻¹, $5\cdot0$ mm and $2\cdot0$ K respectively. Assuming that $H = 4\cdot0$ W m⁻², $T_S = 33$ °C, $T_E = 20$ °C and $e = 1\cdot0$ calculate the rate of heat energy loss due to evaporation, conduction, convection and radiation, and determine the percentage contribution which each transfer mechanism makes to the total rate of heat loss. Comment on the values obtained.

9.10   Show that the units of surface energy (J m⁻²) are the same as those of surface tension (N m⁻¹).

9.11   Which liquids given in Table 9.6 wet (a) paraffin wax and (b) polytetrafluorethylene (PTFE)?

9.12   The contact angle of a solid – liquid combination can be determined by carefully placing a drop of liquid on the surface of the solid and measuring the angle θ (Fig. 9.6b) using an optical goniometer. The following data were obtained for a series of liquids of different surface tension, γ_{LV}, on a bulk sample of stearamide. Determine the critical surface tension γ_c of stearamide by plotting $\cos \theta$ as a function of γ_{LV}. Is this material wet by (a) a soap solution, (b) olive oil?

$\gamma_{LV}/10^{-3}\,N\,m^{-1}$	35	40	48	52	58
$\theta/degree$	35	49	64	73	82

9.13   What is the maximum force experience by a platinum ring of radius 20·0 mm when it is pulled through the surface of water? What mass produces the same force?

9.14   Show, by considering a vertical column of liquid of density ρ (kg m^{-3}), radius R and height H, that the pressure P at a distance H below the surface of a liquid is $P = P_A + H\rho g$ where P_A is the atmospheric pressure and g the acceleration due to gravity. To what depth can the bug Aphelocheirus swim before the water pressure is equal to the maximum pressure that the surface forces can sustain? What happens if the bug goes deeper?

10

Mechanical Properties of Solids and Liquids

In this chapter we discuss two important functional classes of material: structural materials such as bone, tendon and cell wall tissue which transmit forces and contain biological fluids, and transport materials such as blood which move essential substances from one place to another.

10.1 Stress, strain and elastic modulus

Consider a piece of material in the form of a bar of length L_0 and cross-sectional area A_0 as shown in Fig. 10.1a. When an external force F_{ext} is applied to the ends of the bar the atoms or molecules in the bar are displaced in the direction of the applied force so that the length of the bar increases and the width b and thickness d decrease. A single layer of atoms along the axis of the bar is shown in Fig. 10.1b. The atoms move apart until the internal force F_{int} due to the mutual attraction between the atoms is equal to F_{ext}. At equilibrium the internal force across any section XX is equal in size but opposite in direction to the external force, and the length L of the bar then remains constant. The force that exists within the material is described by the stress $\sigma (N\ m^{-2})$ which is defined by the equation

$$\sigma = F/A_0 \tag{10.1}$$

where $F(N)$ is the applied force and $A_0(m^2)$ is the cross-sectional area of the undeformed bar. The equilibrium deformation is described by the strain ϵ which is the elongation $\Delta L = L - L_0$ divided by the initial length L_0

$$\epsilon = \Delta L/L_0 \tag{10.2}$$

Both ΔL and L_0 are measured in metres so that strain is a dimensionless quantity.

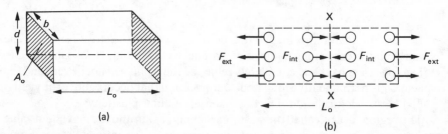

Figure 10.1 (a) An undeformed bar with rectangular cross section. (b) The internal (interatomic) forces F_{int} that oppose the external force F_{ext}.

For a material in which the atoms are arranged in a perfect array the specimen will break when the force across the cross-section is larger than the combined strength of all the atomic bonds in that cross-section. In practice no material has a perfect structure and a specimen often contains impurities, internal voids and surface defects. The stress close to these irregularities is larger than the average stress within the specimen and small cracks are formed at these points. Fracture occurs because of the propagation of one of these cracks through the specimen. The breaking stress (or strength), σ_B, is usually an order of magnitude, or more, less than the theoretical breaking stress.

When the stress in a material increases, the strain increases. There is no unique relationship between stress and strain, different materials behave in different ways. A generalized stress–strain diagram is shown in Fig. 10.2. At low strains most materials exhibit a region, OP, in which the stress increases in proportion to strain so that the σ–ϵ diagram is a straight line. This is called Hookean behaviour because the material obeys Hooke's law (Equ. 2.6). P is the proportional limit. For strains up to ϵ_E the stress–strain behaviour is reversible and when the stress is removed the strain will decrease along the line EPO. Materials which exhibit this reversible behaviour are called elastic and the point E is the elastic limit. Often the points P and E are close together. At higher strains the stress passes through a maximum, called the upper yield point, Y_1. The stress at which this occurs is the yield stress σ_Y. For strains larger than ϵ_Y, the stress decreases until the lower yield point Y_2 is reached. It then increases with increasing strain. A material which exhibits a region $Y_1 Y_2$ is called ductile.

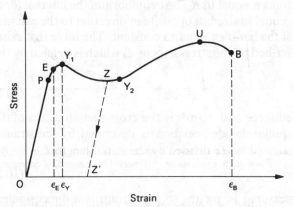

Figure 10.2 A generalized stress–strain diagram. P, proportional limit; E, elastic limit; Y_1, upper yield p¢oind; Y_2, lower yield point; B, break.

If a ductile material is loaded to a point Z and then the stress removed, the strain will decrease along a line, ZZ′, parallel to the initial, Hookean, line OP. When the stress has been completely removed the strain will not be zero and the specimen is permanently elongated. Deformation that occurs above the yield is called plastic deformation. Ductile materials exhibit plasticity.

The region Y_2U is called the strain-hardening region and the maximum value of the stress, σ_U, which occurs at the point U, is the ultimate stress. Some materials do not exhibit a marked yield behaviour but are ductile, i.e. it is difficult to locate the point Y_1 and hence σ_Y. In these cases a line is drawn parallel to OP off-set from the origin along the ϵ-axis by an amount $0\cdot002$ (i.e. $0\cdot2\%$). The point where this line intersects the stress–strain diagram is taken as the yield point. The yield stress determined in this way is called the $0\cdot2\%$ off-set yield stress (see Example 10.2).

Materials can be divided into different classes according to their mechanical behaviour. The σ–ϵ diagrams for three materials which fall into different, fairly well defined, classes are given in Fig. 10.3. Brittle materials, such as window glass, have a breaking strain which is less than the proportional limit so they are Hookean. The σ–ϵ diagram is reversible so they are also elastic. Ductile materials, such as mild steel, are Hookean and elastic at small strains but yield at strains about $0\cdot01$. Above the yield point strain-hardening occurs and the stress increases with increasing strain until the ultimate stress is reached. Ductile materials often have high strength and intermediate values of breaking strain. If the specimen is unloaded before break occurs it is permanently elongated. Rubbery materials, which are more correctly called elastomers, have a very non-linear σ–ϵ diagram but are approximately elastic. The breaking stress is usually low and the breaking strain high. The value of ϵ_B can vary widely according to the composition of the material. Some elastomers can be elongated to ten times their initial length before they break.

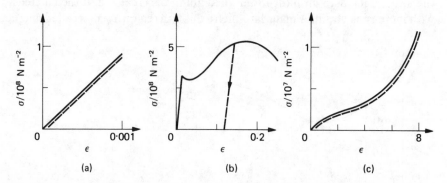

Figure 10.3 Stress-strain diagrams for different classes of material. (a) Silica glass, brittle. (b) Annealed mild steel, ductile. (c) Natural rubber vulcanizate, rubbery. (The unloading curves are indicated by the broken lines.)

Materials can also be classified as being tough or brittle. Toughness is a measure of the energy required to break a material and is related to the area under the σ–ϵ diagram (see Fig. 2.4). Ductile materials are usually tough whereas glassy materials are brittle. Although brittle materials may have high strength their breaking strain is so small that the work done on the specimen is small. Brittle materials are very sensitive to impurities and surface flaws and are easily broken by a sudden impact. Ductile materials withstand impact loading better than brittle materials, most of the energy being dissipated in the plastic deformation and appearing in the form of heat energy.

The type of deformation shown in Fig. 10.1 is called tensile deformation and σ is the tensile stress. Two other types of deformation, shear and bulk, are shown in Fig. 10.4. In shear the force is applied tangential to the surface. An initially rectangular volume element deforms as shown. The deformation is described by the value of the angular distortion γ. For small distortions the shear strain γ is approximately equal to s/h. Volume deformation is caused by a change in pressure. The volume strain is the change in volume, ΔV, divided by the intitial volume, V_0, of the element. The parameters used to describe the three types of deformation are summarized in Table 10.1.

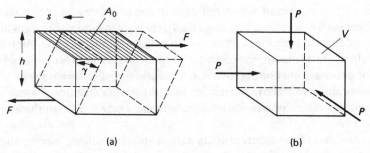

(a)   (b)

Figure 10.4   (a) Shear deformation. (b) Volume (bulk) deformation.

The elastic modulus of a material is defined as the ratio of the stress to strain. For materials which are Hookean (Fig. 10.3a) or part of the response is Hookean (Fig. 10.3b) the modulus is the slope of the straight line part of the σ–ϵ diagram. An elastic modulus can be defined for each type of deformation. The equations relating stress to strain are given in Table 10.1. E is called the tensile (or Young's) modulus, G the shear modulus and K the bulk modulus.

Table 10.1   The three types of stress and strain and the stress–strain equations.

Type	Stress (N m^{-2})	Strain	Stress–strain equation
Tensile	$\sigma = F/A_0$	$\epsilon = \Delta L/L_0$	$\sigma = E\epsilon$
Shear	$\tau = F/A_0$	$\gamma = s/h$	$\tau = G\gamma$
Volume	P	$\Delta V/V_0$	$P = -K\Delta V/V_0$

These equations can be used to calculate the elastic deformation of a structure due to an applied force, or they can be used to determine the modulus from the measured stress–strain diagram.

Example 10.1

A $1\cdot0$ kg mass is attached to the free end of a parallel bundle, $0\cdot56$ mm radius and 10 mm long, of collagen fibres. The other end of the bundle is attached to a rigid support. If the tensile modulus of the bundle is $E = 5\cdot0 \times 10^8\,\text{N m}^{-2}$ what is the elongation?

The mass M exerts a force $F = Mg$ on the bundle. From Table 10.1 and equations 10.1 and 10.2, $\epsilon = \Delta L/L_0 = \sigma/E = F/A_0E$ so that $\Delta L = MgL_0/A_0E =$

$MgL_0/\pi r^2 E$. Converting all quantities to SI units

$$\Delta L = \frac{1\cdot0 \text{ kg} \times 9\cdot8 \text{ m s}^{-2} \times 10 \times 10^{-3} \text{ m}}{3\cdot14 \times (0\cdot56 \times 10^{-3} \text{ m})^2 \times 5\cdot0 \times 10^8 \text{N m}^{-2}}$$

$$\Delta L = 0\cdot2 \text{ mm}$$

10.2   The relationship between structure and properties

The mechanical properties of a material depend upon the strength of the inter-molecular forces (the chemical bonding, §3.4), the way in which the atoms are arranged in relation to each other (the physical structure) and the imperfections in the structure. As far as the mechanical properties are concerned there are two types of bond: strong bonds (covalent, ionic and metallic) and weak bonds (van der Waal and hydrogen). Atoms are joined together by strong bonds to form molecules. Molecules usually interact with each other through the weak bonds.

There are two extremes of structure, crystalline and amorphous. In crystalline solids the atoms or molecules are arranged in a regular way to form a three-dimensional array called a crystal structure (see Fig. 3.9). The term amorphous is usually applied to materials which do not exhibit long-range order and includes gases, liquids and glasses. (It should be remembered that liquids possess short-range order and have densities of the same order of magnitude as those of solids.)

The equilibrium atomic separation in a crystalline solid is approximately the same as the wavelength of X-radiation (§4.8) so that when X-rays are reflected from a crystal, constructive and destructive interference takes place. This is called X-ray diffraction. Consider X-radiation of wavelength λ incident on two planes of atoms in a crystalline solid as shown in Fig. 10.5a. A small part of the incident radiation is scattered by the atoms in much the same way as light waves are reflected from the top and bottom surfaces of a thin transparent film (Fig. 7.12). The ray R_2 reflected from the lower atomic plane travels a distance $\delta = 2d \sin \theta$ further than the ray R_1 reflected from the top plane. If the path difference δ is an integral number of whole wavelengths there will be constructive interference, i.e.

$$m\lambda = 2d \sin \theta \qquad m = 1, 2, 3, \ldots \tag{10.3}$$

This is called Bragg's equation. The X-radiation reflected by a crystal can be detected using a photographic film. For a single crystal the diffraction pattern consists of a series of spots spaced symmetrically about the axis of the incident beam as shown in Fig. 10.5b. If the specimen is made up of an agglomeration of small randomly oriented crystals the pattern is a series of concentric circles. The interplanar spacing d can be determined from the size of the circles in the diffraction pattern.

In atomic and ionic crystals the atoms or ions are bonded together by the strong bonds: e.g. diamond (covalent), sodium chloride (ionic) and metals (metallic). The modulus of crystals is usually high because all the atomic bonds act together to resist the applied stress. Covalent and ionically bonded crystals fracture at small elongations by the rapid propagation of a crack in a direction

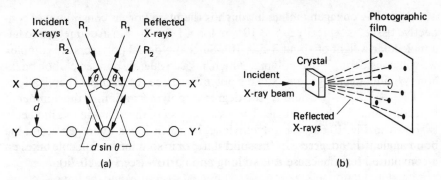

Figure 10.5 (a) The reflection of X-rays from two atomic planes in a crystal. (b) The X-ray diffraction pattern of a single crystal.

perpendicular to the direction of the applied stress. This is characteristic of brittle materials. Although metals exist in a crystalline form, in the same way that diamond and sodium chloride are crystalline, metals are often ductile rather than brittle. The reason for this is the simplicity of their crystal structure. In metals the atoms are arranged in well defined planes. The interatomic spacing in some planes is smaller than in others, i.e. they have a higher packing density. Two such planes, XX' and YY', are shown in Fig. 10.5a. The separation, d, between the close-packed planes is larger than the interatomic separation within the planes and it is relatively easy for these close-packed planes to slip past each other under the action of a shear stress. The slip is non-recoverable (i.e. non-elastic) and gives rise to the ductile behaviour exhibited by many metals. An applied tensile stress sets up shear stresses within a specimen which have a maximum value in a direction 45 degrees to the direction of the tensile stress. Fracture takes place by slip in a direction close to the direction of maximum shear stress so that with ductile failure the fracture surface is inclined to the axis of the specimen.

Glasses are amorphous solids and are sometimes called supercooled liquids. However, not all liquids form a glass. At (relatively) high temperatures glassy materials exist as liquids and there is a considerable amount of space, called free volume, between the molecules. As the temperature is decreased the thermal energy of the molecules decreases and there is a contraction in the free volume so that the packing density increases. This results in an increase in viscosity (§ 10.5). The viscosity increases continuously with decreasing temperature, and the abrupt change exhibited by crystalline materials at the crystal melting temperature does not occur. At (relatively) low temperatures the viscosity is so high that the material behaves as a brittle solid. It has high modulus and low breaking strain. By high and low temperatures we mean relative to our normal experience. For example, at room temperature Pyrex glass is a solid whereas at high temperature it is a liquid. At room temperature unvulcanized natural rubber is a viscous liquid whereas at -100 degree Celsius it is a glass.

Most living tissue is made up of very long, high molecular weight, molecules called polymers (e.g. proteins in animals and cellulose in plants) which consist of a covalently bonded backbone chain to which pendant side groups are attached. The collagen molecule which is the main fibrous constituent of connective tissue, is typically $3 \cdot 5 \times 10^{-7}$ m long, $1 \cdot 4 \times 10^{-9}$ m in diameter and has a molecular weight of about $3 \cdot 5 \times 10^5$. Since biological polymers are complex we will consider a simple polymer which has a covalently bonded carbon backbone chain and side groups X. This polymer can be represented chemically as $[- CX_2 - CX_2 -]_N$ where N, the degree of polymerization, is the number of repeating units in the molecule. The arrangement of the atoms in this molecule is shown in Fig. 10.6a. For polyethylene, in which X is a hydrogen atom, the bond angle is 109 degrees. In the solid state, or in solution, a molecule takes up a convoluted form because it is so long and narrow (see Fig. 10.7a).

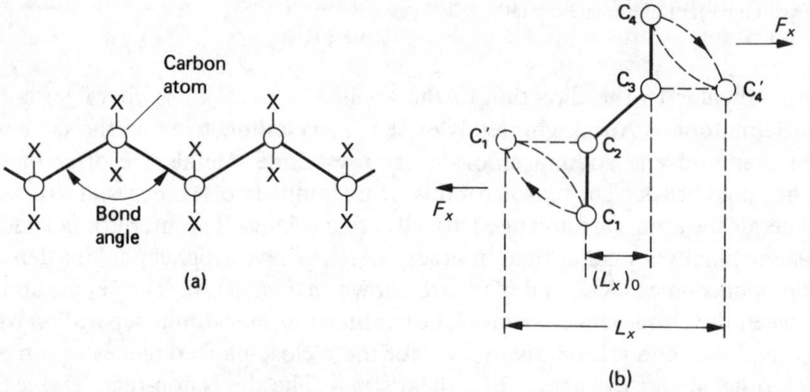

Figure 10.6 (a) A small segment of a polymer molecule with a carbon backbone chain. (b) The elongation of a segment by bond rotation due to a force F_x.

Deformation of the molecule, and hence the material it comprises, under the action of a force can take place in two ways: by bond-angle opening and by the rotation of single carbon atoms, or segments of the molecule, about the backbone chain. A segment of the backbone chain is shown in Fig. 10.6b. The initial length of the segment, between carbon atoms C_1 and C_4, resolved in the x-direction is $(L_x)_0$. When the segment experiences a force F_x, in the x-direction, atoms C_1 and C_4 can rotate into positions C_1' and C_4', the bond angle remaining the same. The length of the segment in the x-direction is now L_x, which is larger than $(L_x)_0$. Since the force required for bond rotation is less than that for bond-angle opening, deformation is usually due to segmental rotation and the material is rubbery. However, for rotation to occur there must be sufficient space, i.e. free volume, for the atoms, or segments of molecule, to rotate into. If the pendant side groups are large, segmental rotation is restricted and the material is not easily deformed. For example if polyethylene existed in amorphous form at room temperature it would be rubbery. (In practice polyethylene is not amorphous but semi-crystalline and because of this it is ductile, as explained below.) However, polystyrene, which has large pendant benzene rings at regular intervals along the chain, is rigid because there is insufficient free volume for segmental rotation to occur. In this case deformation is through bond-angle opening. Any factor that decreases the free volume, such as the lowering of the temperature, causes an increase in the rigidity of the material. The temperature at which flexible elastomers transform to brittle glasses is called the glass transition temperature.

When a large number of polymer molecules combine to form a piece of material the molecules become entangled. Two molecules AA′ and BB′ are shown in Fig. 10.7a. Interaction between the molecules occurs through the weak bonds so that when A is pulled to the left and B′ to the right there is an initial elastic deformation, due to segmental rotation, and then the molecules slip over each other. These materials cannot be used for structural purposes, i.e. to support a load, because of this continuous slipping of the molecules. Chain slippage can be stopped by joining the molecules together at various points using strong (covalent) bonds. This is called crosslinking and can be done chemically or by means of ionizing radiation (see § 12.1). The process by which rubber is crosslinked is called vulcanization.

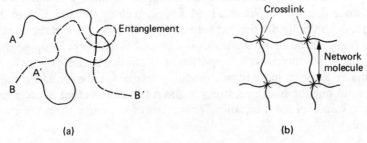

(a)   (b)

Figure 10.7   (a) Two convoluted, and entangled, polymer molecules. (b) A molecular network formed by crosslinking.

A crosslinked elastomer is shown schematically in Fig. 10.7b. The kinetic theory of rubber elasticity shows that the initial tensile modulus E of an elastomer, i.e. the slope of the $\sigma - \epsilon$ diagram at ϵ tending to zero is $E = 3\rho RT/M_c$ where ρ(kg m^{-3}) is the density, $R(= 8 \cdot 31 \times 10^3$ J kmol^{-1} K^{-1}) is the gas constant and T(K) is the absolute temperature. M_c is the molecular weight of a network molecule, i.e. the molecule between crosslinks. The modulus increases as M_c decreases. It is possible to convert a flexible rubbery material into a hard, tough, material by introducing a large number of cross-links.

When the side groups attached to the backbone chain are small, and arranged in a regular way, segments of different molecules are able to combine and form small crystalline regions, called crystallites. These crystallites are embedded in a matrix of amorphous polymer (Fig. 10.8a). Intermolecular bonding in these molecular crystals is through the weak bonds, e.g. polyethylene (van der Waals) and cellulose (hydrogen). Different molecules can be incorporated into the same crystalline region so that the crystallites are not only regions of high modulus, they effectively tie the molecules together and act as crosslinks. Most synthetic and biological fibres (e.g. nylon, collagen, keratin, cellulose) are of this type and are called semi-crystalline. They are ductile and have intermediate values of modulus, the value depending on the degree of crystallinity, i.e. the volume fraction of crystalline material.

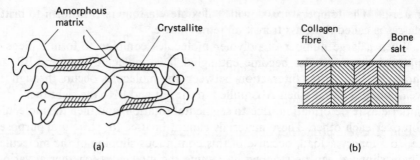

Figure 10.8 (a) A semi-crystalline polymer. (b) A composite material (bone).

Bone is a composite material consisting of a parallel array of collagen fibres firmly attached to a matrix of small inorganic crystals, called the bone salt, as shown in Fig. 10.8b. When a load is applied along the direction of the fibres the stress is initially distributed between the components and the tensile modulus is high. At quite small strains the crystal matrix breaks up so that yield occurs and then the stress is supported by the fibres. Because of this bone is a tough material which has a tensile modulus between that of collagen and a crystalline solid but the breaking stress is about the same as that of collagen. Approximate values of the mechanical properties of some materials are given in Table 10.2.

Table 10.2 The mechanical properties and structure of some materials.

Material	$E/\text{N m}^{-2}$	$\sigma_B/\text{N m}^{-2}$	Structure
Elastin	6×10^5		A crosslinked amorphous polymer network; rubbery. Elastic protein found in the connective tissue of vertebrates.
Natural rubber vulcanizate	5×10^5	1×10^7	A lightly crosslinked amorphous polymer network; rubbery. Obtained from the tree Hevea Braziliensis.
Resilin	2×10^6	3×10^6	A crosslinked amorphous polymer network; rubbery. Found in the thorax of insects.
Collagen	1×10^9	1×10^8	A semi-crystalline polymer; ductile. Found in tendons and ligaments.
Bone	1×10^{10}	1×10^8	A composite of a semi-crystalline polymer and an inorganic salt; tough. A major structural material.
Mild steel	1×10^{11}	4×10^8	A randomly oriented agglomerate of small crystals; tough. A structural material.

10.3 Measurement of the stress–strain behaviour

The mechanical properties of a structure can be determined in two ways. If the structure has simple geometry (e.g. artery wall) the properties of a specimen of the material can be measured and the properties of the structure calculated using these values. If the structure has complex geometry (e.g. bone) it is easiest to apply a force to the complete structure and measure the resulting deformation. The first type of measurement gives the properties of the material and the second gives the properties of a specific structure.

The simplest arrangement for determining the tensile force–elongation behaviour of a structure, or a specimen, is to fix one end of the specimen to a rigid support and hang a mass M on the free end. The mass exerts a force Mg on the specimen and the specimen elongates. The elongation can be measured using a millimetre scale and the stress, $\sigma = Mg/A_0$, and the strain, $\epsilon = \Delta L/L_0 = (L - L_0)/L_0$, determined. If the specimen has high stiffness ($K = F/\Delta L$, see Equ. 2.6) a displacement transducer, such as an LVDT (§6.3), must be used to measure the elongation. This method has limited application because it takes a long time to determine the behaviour over a wide range of applied forces, it does not give a permanent record of the force–elongation behaviour and the sample cannot be elongated at different rates.

A schematic diagram of an apparatus that is widely used for the measurement of the mechanical properties of materials is shown in Fig. 10.9. It consists of a rigid frame and a crosshead which is driven upwards or downwards by means of two rotating threaded rods. The specimen is clamped in grips. The upper grip is attached to a force transducer, called a load cell, and the lower grip is attached to the crosshead. When the crosshead is driven downwards the upper grip remains stationary and the specimen is elongated. The elongation of the specimen is equal to the movement of the crosshead. The load cell produces

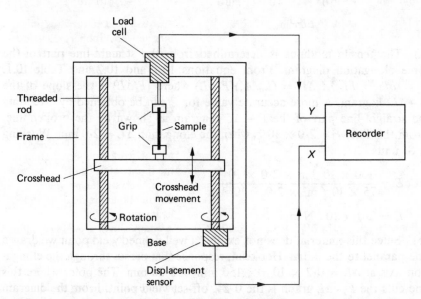

Figure 10.9 A schematic diagram of an apparatus used for the measurement of the mechanical properties of materials (tensile mode).

an electrical signal proportional to the force experienced by the specimen. This signal is conditioned and then fed to the vertical, Y, input of a chart recorder. The movement of the crosshead is monitored, and a signal proportional to this movement is fed to the horizontal, X, input of the recorder. The recorder displays automatically the force–elongation diagram for the specimen. The load cell is calibrated by removing the specimen and attaching known masses. This apparatus allows a wide variety of loading conditions to be investigated. Examples of its applications include the study of different rates of elongation, periodically varying elongations and the variation of force with time at fixed elongations.

Example 10.2

The tensile force–elongation diagram for a strip of bovine femur is shown below. If the undeformed length $L_0 = 150$ mm and cross-sectional area $A_0 = 75$ mm^2 what is: (a) the tensile modulus E; (b) the yield strength σ_Y; (c) the breaking strength σ_B; and (d) the breaking strain ε_B?

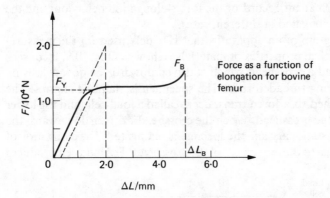

Force as a function of elongation for bovine femur

(a) The tensile modulus is determined from the straight-line part of the force–elongation diagram. From equations 10.1 and 10.2 and Table 10.1, $E = \sigma/\varepsilon = FL_0/A_0\Delta L = (L_0/A_0)(F/\Delta L)$ where $(F/\Delta L)$ is the slope of the $F - \Delta L$ diagram. A more accurate value for E can be obtained by extending the straight-line part of the $F - \Delta L$ diagram as shown by the broken line. From this line, $F = 2\cdot0 \times 10^4$N when the elongation $\Delta L = 2\cdot0$ mm. Working in SI units

$$E = \frac{150 \times 10^{-3}\ \text{m} \times 2\cdot0 \times 10^4\ \text{N}}{75 \times 10^{-6}\ \text{m}^2 \times 2\cdot0 \times 10^{-3}\ \text{m}}$$

$$E = 2\cdot0 \times 10^{10}\ \text{N m}^{-2}$$

(b) Since this material does not exhibit a well defined yield point we draw a line parallel to the initial, Hookean, response that passes through the elongation axis at $\Delta L = 0\cdot2 \times 10^{-2} \times 150$ mm $= 0\cdot30$ mm. The point where this line cuts the $F - \Delta L$ graph is the $0\cdot2\%$ off-set yield point. From the diagram $F_Y = 1\cdot2 \times 10^4$N so $\sigma_Y = F_Y/A_0 = (1\cdot2 \times 10^4\text{N})/(75 \times 10^{-6}\ \text{m}^2) = 1\cdot6 \times 10^8\text{N m}^{-2}$.

(c) The breaking strength $\sigma_B = F_B/A_0 = (1\cdot5 \times 10^4 N)/(75 \times 10^{-6} m^2) = 2\cdot0 \times 10^8 N\,m^{-2}$.

(d) The breaking elongation $\Delta L_B = 5\cdot0$ mm, so the breaking strain $\varepsilon_B = \Delta L_B/L_0 = (5\cdot0 \times 10^{-3} m)/(150 \times 10^{-3} m) = 0\cdot033 = 3\cdot3\%$.

10.4 Stored elastic energy

It was shown in Chapter 2 that when a Hookean spring is extended by an amount ΔL, work $W = (1/2)K \Delta L^2$ is done on the spring and that this work is stored in the spring as potential energy. This is called the stored elastic energy. For a specimen of Hookean material with tensile modulus E, initial length L_0 and area A_0 the stiffness $K = E(A_0/L_0)$. Since $\Delta L = \varepsilon L_0$ the stored elastic energy in a Hookean material extended by an amount ΔL is $W = (1/2)(E A_0/L_0)(\varepsilon L_0)^2$ or

$$W = E\varepsilon^2 V_0/2 = \sigma\varepsilon V_0/2 \tag{10.4}$$

where $V_0 = A_0 L_0$ is the undeformed volume of the specimen. The second equality is obtained by substituting $\sigma = \varepsilon E$.

In practice most materials are not completely elastic and only part of the work done is stored as elastic potential energy. The remainder is dissipated in the form of heat and the temperature of these materials increases when they are deformed. The energy storage properties of a material are described by the resilience R, which is defined as

$$R = W_s/W \tag{10.5}$$

W_s is the stored elastic energy (which can be recovered from the specimen) and W is the work that is done on the specimen (Equ. 10.4). The resilience is related to the logarithmic decrement Φ (§ 2.3) by $1/R = 1 + \Phi/\pi$. Energy storage is important in the flight and jump mechanism of some insects.

The jumping ability of the flea and the locust (§1.5) cannot be explained in terms of simple muscle action. It can be accounted for by a mechanism which involves stored elastic energy. With fleas most of the force required for jumping is transmitted to the ground by the large hind legs. The legs are connected to the body by a ball and socket joint, the space between the ball and socket being filled with resilin, which is a high resilience rubbery material. The arrangement is shown schematically in Fig. 10.10. The flea prepares itself for flight by pulling its body downwards using the muscles shown. This compresses the pad of resilin in the joint. When the compression is completed two catches latch onto the leg and hold the resilin in its compressed state and the compressor muscles relax. When the catches are removed the elastic energy stored in the resilin pad is released very rapidly and the body of the flea is accelerated in the upwards direction. (This principle is used in the elastic catapult. The rubber is stretched slowly by means of muscle action. The energy is stored in the rubber and is suddenly transmitted to the missile when the rubber is released.)

From equations 10.4 and 10.5 the energy stored in the resilin is $W_s = RE \times \varepsilon^2 V_0/2$. Approximate values for the parameters are $E = 2\cdot0 \times 10^6 N\,m^{-2}$,

$R = 0.97$, $V_0 = 1.0 \times 10^{-12}$ m³. Taking the strain ε as 0.30 (which is reasonable for a rubber pad in compression)

$$W_s = \frac{0.97 \times 2.0 \times 10^6 \text{N m}^{-2} \times (0.30)^2 \times 1.0 \times 10^{-12} \text{ m}^3}{2}$$

$$W_s = 9.0 \times 10^{-8} \text{ J}$$

This stored elastic energy is converted into kinetic energy of motion. From Equ. 1.26 the jump height $H = W_s/Mg$ where M is the mass of the flea, which is typically 0.20×10^{-6} kg. Thus

$$H = \frac{9.0 \times 10^{-8} \text{ J}}{2.0 \times 10^{-7} \text{ kg} \times 9.8 \text{ m s}^{-2}} = 4.6 \times 10^{-2} \text{ m} = 46 \text{ mm}$$

This is more than one half of the jump height of the typical flea. Additional energy for the jump is supplied by other mechanisms.

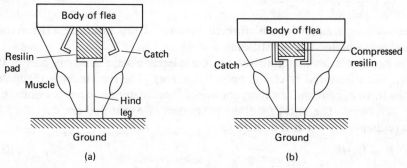

Figure 10.10 The hind leg of a flea: (a) resting; (b) cocked for jump.

10.5 Viscous flow

In liquids and gases the thermal energy of the molecules is large so that the molecules are able to break the intermolecular bonds and they are free to move within the system. Normally this movement is completely random in direction. When a shear stress is applied to a liquid or a gas the movement of the molecules is biased in the direction of the stress (Fig. 10.11a) and there is a net transport of material from one location to another. This is called flow. The flow of a fluid is opposed by internal friction forces due to the weak intermolecular bonds.

Consider a layer of liquid, thickness h, flowing over a stationary surface as shown in Fig. 10.11b. The movement of the molecules is caused by a shear stress τ applied to the top plane of the liquid layer. The molecules in contact with the fixed surface are stationary. In a small time Δt the section AB moves a distance Δs to A'B' and the volume element is deformed by an amount $\Delta \gamma$. The shear stress τ(N m^{-2}) is proportional to the time rate of change of the shear strain $\Delta \gamma / \Delta t$ (s^{-1}), the constant of proportionality, η, being called the coefficient of viscosity:

$$\tau = \eta \frac{\Delta \gamma}{\Delta t} \tag{10.6}$$

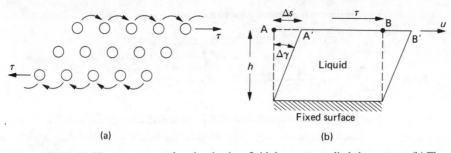

Figure 10.11 (a) The movement of molecules in a fluid due to an applied shear stress. (b) The flow of liquid over a fixed surface.

The strain rate $\Delta\gamma/\Delta t = \Delta(s/h)/\Delta t = (\Delta s/\Delta t)/h = u/h$ where u is the particle velocity in the plane which is distance h from the fixed surface. u/h is called the velocity gradient.

The coefficient of viscosity is a material property in the same way that the elastic modulus is a material property. Equ. 10.6 is called Newton's law of viscous flow. It is analogous to the equation $\tau = G\gamma$ which defines the shear modulus of a Hookean elastic material. Approximate values for the coefficient of viscosity of several fluids at room temperature are given in Table 10.3.

Table 10.3 Coefficients of viscosity at room temperature.

Fluid	$\eta/\mathrm{N\,m^{-2}s}$	Fluid	$\eta/\mathrm{N\,m^{-2}s}$
Air	$1\cdot8 \times 10^{-7}$	Caster oil	$0\cdot98$
Water	$1\cdot0 \times 10^{-3}$	Synovial fluid	$1\cdot0$
Blood	$3\cdot5 \times 10^{-3}$	Glycerol	$1\cdot5$

For a liquid to flow along a tube it is necessary to apply a pressure to one end of the tube to overcome the viscous flow stress τ which opposes the flow. Consider a liquid with viscosity η flowing from left to right in a tube of length L and radius a (Fig. 10.12). If the pressure difference between the two ends of the tube is P the speed u of the molecules at a distance r from the axis is

$$u = \frac{P}{4L\eta}(a^2 - r^2) \tag{10.7}$$

The molecular velocity u is a maximum at the centre of the tube ($r = 0$) and is zero at the walls ($r = a$). The volume flow velocity $Q(\mathrm{m^3\,s^{-1}})$, which is the volume of liquid flowing along the tube in one second, is given by Poiseuille's equation:

$$Q = \frac{\pi}{8}\frac{Pa^4}{L\eta} \tag{10.8}$$

It is possible to define the flow resistance of a system in the same way as we defined the electrical resistance of a circuit element using Ohm's law, $V = IR$ (§ 5.1). For flow the equivalent equation is

$$P = QR_\mathrm{F} \tag{10.9}$$

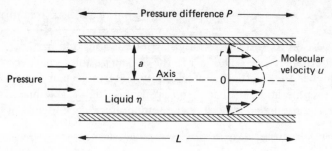

Figure 10.12   The flow of liquid along a pipe.

where R_F is the flow resistance. For a circular section tube $R_F = 8\eta L/\pi a^4$. When several tubes are connected in series the total resistance $R_F' = R_1 + R_2 + R_3 \ldots$ and when in parallel $1/R_F' = 1/R_1 + 1/R_2 + 1/R_3 \ldots$.

Example 10.3

(a) What is the total flow resistance of the three parallel arteries in the calf which have radius 1·0 mm and length 200 mm? (b) If the volume flow velocity of blood through these arteries is $1·7 \times 10^{-6}$ m³ s⁻¹ what is the pressure drop across the arteries?

(a)   For a single artery $R_F = 8\eta L/\pi a^4$ and the viscosity of blood is approximately $3·5 \times 10^{-3}$ N m⁻² s. Consequently

$$R_F = \frac{8 \times 3·5 \times 10^{-3}\,\text{N m}^{-2}\,\text{s} \times 200 \times 10^{-3}\,\text{m}}{\pi \times (1·0 \times 10^{-3}\,\text{m})^4}$$

$$R_F = 1·8 \times 10^9\,\text{N m}^{-5}\,\text{s}$$

There are three arteries in parallel so that the total resistance $1/R_F' = 1/R_F + 1/R_F + 1/R_F$ or $R_F' = R_F/3$ so that

$$R_F' = 6·0 \times 10^8\,\text{N m}^{-5}\,\text{s}$$

(b)   The pressure drop across the arteries is $P = QR_F'$ so that

$$P = 1·7 \times 10^{-6}\,\text{m}^3\,\text{s}^{-1} \times 6·0 \times 10^8\,\text{N m}^{-5}\,\text{s}$$

$$P = 10 \times 10^2\,\text{N m}^{-2} = 7·5\,\text{mm Hg}$$

For Newton's law and Poiseuille's equation to be valid the molecules of the fluid must flow along the tube in a uniform manner so that they are always the same distance from the axis. This is called streamlined flow. When the flow velocity is increased beyond a critical value, Q_c, the flow becomes unstable and vortices are formed in the flowing fluid. This is called turbulent flow. The onset of turbulence is governed by a combination of four parameters: the density ρ, Q, a and η. When $\rho Q/\pi a \eta$ is less than a critical value, called Reynold's number Re, the flow is streamlined and when it is larger than Re the flow is turbulent. For a pipe with circular cross section $Re = Q_c \rho/\pi a \eta$ and the value of Re is approximately 2×10^3. The pressure difference necessary to produce a given flow velocity with turbulent flow is larger than if the flow is streamlined (Fig. 10.13a).

Viscous flow resistance is due to molecular interactions and may be considered as an internal friction. The larger the attractive forces between the molecules the higher the coefficient of viscosity. Intermolecular forces are very sensitive to temperature so that the viscosity of a liquid is temperature dependent. The equation that describes the temperature dependence of the viscosity

$$\eta_T = A \, e^{E/RT} \tag{10.10}$$

is of the same form as the equation that describes the rate of a chemical reaction. The pre-exponential factor A is approximately a constant, R is the gas constant and E (J kmol^{-1}) is the activation energy for viscous flow. The viscosity of a liquid decreases with increasing temperature.

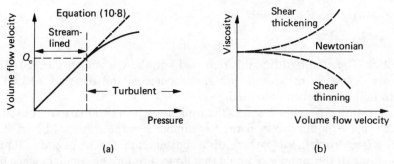

Figure 10.13 (a) The P–Q relationship for streamlined and turbulent flow. (b) Viscosity–flow velocity relationships for Newtonian and non-Newtonian fluids.

So far we have assumed that the viscosity is a property of the material which is independent of shear rate (or volume flow velocity Q). This is true for pure liquids and some solutions but it is not true for suspensions or for solutions of complex molecules such as polymers. These liquids are called non-Newtonian. Many different types of non-Newtonian behaviour occur. Two simple examples are given in Fig. 10.13b. With shear thickening (e.g. a concentrated suspension of corn starch) the viscosity increases with increasing flow rate whereas with shear thinning (e.g. synovial fluid) the viscosity decreases with increasing rate.

10.6 Measurement of the viscosity of liquids

The viscosity of a liquid can be measured using a U-tube (or Ostwald) viscometer. This consists of a glass U-tube with two or three bulbs and a vertical capillary of length L as shown in Fig. 10.14a. The viscometer is calibrated using a standard liquid (such as silicone oil). A predetermined amount of the standard liquid is introduced into the right-hand limb and is then drawn up to the line A on the left-hand limb by means of suction. The liquid is allowed to flow through the capillary and the time t_1 taken for the level to fall to B is

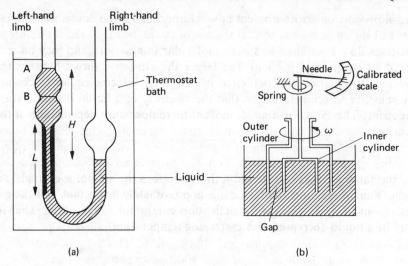

Figure 10.14   (a) A U-tube viscometer. (b) A concentric cylinder viscometer.

measured. The standard liquid is removed and the viscometer cleaned. The unknown liquid is introduced into the viscometer and the time t_2 for the level to fall from A to B measured.

The pressure difference P across the capillary is proportional to the distance H between the liquid levels in the two limbs. For the standard liquid $P = k H \rho_1 g$ (see Problem 9.14) where k is a constant, and for the unknown liquid $P = k H \rho_2 g$. If the amount of liquid that flows through the capillary when the level falls from A to B is q, Equ. 10.8 gives

$$\frac{q}{t_1} = \frac{\pi}{8} \frac{k H \rho_1 g \, a^4}{\eta_1 L} \text{ and } \frac{q}{t_2} = \frac{\pi}{8} \frac{k H \rho_2 g \, a^4}{\eta_2 L}$$

where L is the length and a the radius of the capillary. Since q is the same in each case

$$\frac{\eta_2}{\eta_1} = \frac{t_2 \rho_2}{t_1 \rho_1}$$

If η_1, ρ_1 and ρ_2 are known, the viscosity η_2 of the unknown liquid can be determined. The product η/ρ is called the kinematic coefficient of viscosity. This is not an absolute method, the unknown liquid is compared with a standard liquid.

The U-tube viscometer provides values of the viscosity at low shear rates, $\Delta \gamma / \Delta t$ approximately equal to zero. Sometimes it is necessary to measure the viscosity at high rates. This can be done using a concentric cylinder viscometer which is shown in Fig. 10.14b. When the outer cylinder is rotated at speed ω (radian s^{-1}) the inner cylinder experiences a turning force because of the flow of liquid in the gap between the cylinders. The inner cylinder rotates until the restoring force due to the spring equals the viscous flow force. The equilibrium position of the inner cylinder, which is indicated by the needle and calibrated scale, is proportional to the viscosity of the liquid. The outer cylinder can be rotated at different speeds so that the viscosity can be measured for different shear rates.

10.7 Viscoelasticity

With amorphous polymeric solids most of the deformation is due to the rotation of chain segments about the main chain backbone. This rotation is not completely free because the chain segments collide with and slide over other molecular chains. The rotation is impeded by internal friction forces in the same way as the flow of a simple liquid is impeded by internal friction. If the molecular chains are not connected together by physical crosslinks, or by some other mechanism, the chains can slide over each other and viscous flow takes place. Because of this, polymeric materials exhibit simultaneously properties characteristic of viscous liquids and elastic solids. This type of material is called viscoelastic. Hindered segmental rotation cannot take place instantaneously so that the deformation under the action of an applied force is time dependent. Most load-bearing biological materials (e.g. skin, bone, muscle tissue) are viscoelastic.

Suppose a force F is applied to a specimen of viscoelastic material at time $t = 0$. The deformation of the material is made up of three parts. (i) An instantaneous elastic deformation, ΔL_i, due to bond stretching. This occurs very rapidly but once it has taken place it remains constant, independent of time. (ii) A retarded elastic deformation, ΔL_r, due to the hindered rotation of large segments of the chain backbone. This deformation increases with increasing time but eventually reaches an equilibrium value. (iii) A flow deformation, ΔL_f, which is due to the sliding of complete chains through the material. This deformation increases in proportion to time. The time dependence of the total deformation and the three components are shown in Fig. 10.15. Deformation under the action of a constant force is called creep.

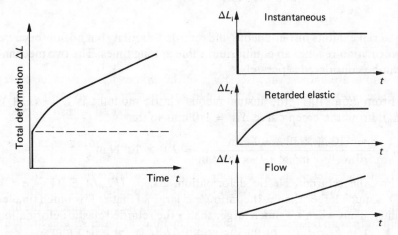

Figure 10.15 The creep of a viscoelastic material. The total deformation ΔL and the three components ΔL_i, ΔL_r and ΔL_f.

The total deformation $\Delta L = \Delta L_i + \Delta L_r + \Delta L_f$ due to a force F applied at $t = 0$ can be described by the equation

$$\Delta L = \frac{FL_0}{A_0} \left(\frac{1}{E_i} + \frac{1}{E_r}(1 - e^{-t/\tau}) + \frac{t}{\eta} \right) \tag{10.11}$$

where L_0 and A_0 are the undeformed length and cross-sectional area of the specimen, E_i is the instantaneous elastic modulus, E_r the retarded elastic modulus and η the flow viscosity. τ is the retardation time of the material (it is similar to a time constant τ, §5.3). The presence of a particular deformation mechanism can be determined from the shape of the creep curve.

Example 10.4

A constant force of 10 N is applied to a piece of tendon of initial length $L_0 = 10$ mm and cross-sectional area $A_0 = 5 \cdot 0 \times 10^{-7}$ m². The creep curve is shown below. What is: (a) the instantaneous elastic modulus; (b) the retarded elastic modulus; and (c) the retardation time for this material?

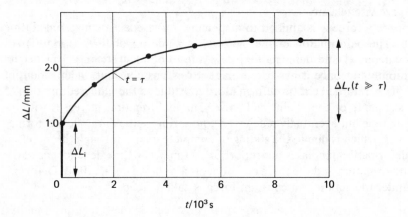

The material exhibits instantaneous and retarded elasticity but no flow because the deformation reaches an equilibrium value at long times. The two mechanisms can be considered separately.

(a) From Equ. 10.11 the instantaneous elastic modulus is $E_i = (F/A_0)/(\Delta L_i/L_0)$. From the creep curve $\Delta L_i = 1 \cdot 0$ mm so that

$$E_i = \frac{10 \text{ N} \times 10 \times 10^{-3} \text{ m}}{5 \cdot 0 \times 10^{-7} \text{ m}^2 \times 1 \cdot 0 \times 10^{-3} \text{ m}} = 2 \cdot 0 \times 10^8 \text{ N m}^{-2}$$

(b) For the retarded elastic deformation, $\Delta L_r = (FL_0/A_0 E_r)(1 - e^{-t/\tau})$. When t is much larger than τ, the ratio t/τ is large so that $e^{-t/\tau}$ is much smaller than unity. Thus when t is much larger than τ the retarded elastic deformation $\Delta L_r(t \gg \tau) = FL_0/A_0 E_r$. From the graph $\Delta L_r(t \gg \tau)$ is $1 \cdot 5$ mm so

$$E_r = \frac{10 \text{ N} \times 10 \times 10^{-3} \text{ m}}{5 \cdot 0 \times 10^{-7} \text{ m}^2 \times 1 \cdot 5 \times 10^{-3} \text{ m}} = 1 \cdot 3 \times 10^8 \text{ N m}^{-2}$$

(c) Considering the retarded elastic deformation, if $t = \tau$ then $\Delta L_r = (FL_0/A_0E_r)(1 - e^{-1})$. Since $e^{-1} = 1/e = 0\cdot37$ and $(FL_0/A_0E_r) = L_r(t \gg \tau)$, from part (b), the retarded elastic deformation when $t = \tau$ is

$$\Delta L_r(t = \tau) = 0\cdot63\,\Delta L_r(t \gg \tau)$$

Since $\Delta L_r(t \gg \tau) = 1\cdot5$ mm, $\Delta L_r(t = \tau) = 0\cdot95$ mm and the total deformation at this time will be $\Delta L = \Delta L_i + \Delta L_r(t = \tau) = 1\cdot0 + 0\cdot95 = 1\cdot95$ mm. From the creep curve the time t when the total deformation is $1\cdot95$ mm is $t = 2 \times 10^3$ s so that the retardation time $\tau = 2 \times 10^3$ s.

Problems

10.1   The tensile stiffness K of a sample is defined by the equation $F = K\Delta L$ and the modulus E of the material by $\sigma = E\varepsilon$. Express K in terms of E for a sample of initial length L_0 and cross-sectional area A_0. Explain the difference between stiffness and modulus.

10.2   What is the network molecular weight of a specimen of resilin which has initial tensile modulus $E = 2\cdot0 \times 10^6$ N m^{-2} and density $\rho = 0\cdot50 \times 10^3$ kg m^{-3} at 27°C? What would be the modulus if M_c were half this value?

10.3   Explain, with the aid of diagrams, what is meant by (a) Hookean, (b) elastic, (c) ductile and (d) non-Hookean (or non-linear) behaviour. Describe the difference between brittle and tough materials, and, using examples from your own experience, discuss the importance of toughness.

10.4   The mechanical properties of materials made up of oriented polymer molecules or oriented fibres depend upon the direction in which the force is applied. This is called mechanical anisotropy. The following data were obtained for specimens of cartilage tissue cut: (a) parallel; and (b) perpendicular to the direction of orientation of the collagen fibres. (i) Draw the tensile stress–strain diagrams. (ii) Determine the tensile moduli from the straight-line part of the σ–ε diagrams. (iii) What is the direction of highest modulus?

Strain	0·00	0·02	0·04	0·06	0·08	0·10	0·12
Stress (a)/10⁶ N m⁻²	0·00	0·50	2·0	5·0	8·0	11·0	14·0
Stress (b)/10⁶ N m⁻²	0·00	0·30	0·50	1·2	2·0	2·8	3·5

10.5   A plastic syringe can withstand a pressure of $2\cdot0 \times 10^4$ N m^{-2}. Explain what happens when a solution with viscosity $2\cdot0 \times 10^{-3}$ N m^{-2} s is forced through a hyperdermic needle of length 20 mm and diameter $0\cdot50$ mm at a rate of $1\cdot0 \times 10^3$ mm^3 s^{-1}.

10.6   The pressure drop P across the various parts of the circulatory system are: (a) arteries, 10 mm Hg; (b) artereoles, 55 mm Hg; (c) capillaries, 20 mm Hg; and (d) veins, 12 mm Hg. If the volume flow velocity Q of the blood is $1\cdot1 \times 10^{-4}$ m^3 s^{-1} calculate in SI units the flow resistance of: (a) the different parts of the system; and (b) the complete system.

10.7   The heart is a pump which produces constant flow velocity Q. What would be the pressure difference across the pump if the blood were replaced by a saline solution which has the same viscosity as water? (See Problem 10.6.)

10.8   What is the maximum particle velocity, u, of blood flowing in an artery of 10 mm diameter at a volume flow rate $Q = 6\cdot6$ litres/min? (Work in SI units.)

10.9   Discuss the significance of turbulence on the behaviour of the circulatory system.

10.10   The following data were obtained for the viscosity η of synovial fluid at different shear rates $\Delta\gamma/\Delta t$. Plot a graph of $\log_{10}\eta$ as a function of $\log_{10}(\Delta\gamma/\Delta t)$ and establish an equation relating η to $\Delta\gamma/\Delta t$. (Hint, if $A = B^X$ then $\log_{10} A = X\log_{10} B$.)

$(\Delta\gamma/\Delta t)/\mathrm{s}^{-1}$	0·1	1	10	100	1000
$\eta/\mathrm{N\,m^{-2}\,s}$	$6\cdot4 \times 10^{-1}$	$1\cdot6 \times 10^{-1}$	$2\cdot5 \times 10^{-2}$	$7\cdot9 \times 10^{-3}$	$1\cdot3 \times 10^{-3}$

10.11   Explain why it is not possible to describe the mechanical properties of many biological tissues simply in terms of an elastic modulus.

10.12   The following data were obtained during a creep experiment on tissue taken from the wall of a sea anemone. The stress was $\sigma_0 = 1 \times 10^4\,\mathrm{N\,m^{-2}}$. (a) Plot a graph of ε as a function of t. (b) What deformation mechanisms are present? (c) Determine the value of E_i, E_r and τ. (d) Calculate using these values the strain ε predicted by Equ. 10.11 for different values of t. Plot the theoretical and experimental values of ϵ as a function of $\log_{10} t$. Comment on the fit between the two curves.

ϵ	0·20	0·30	0·40	0·60
t/s	$3\cdot6 \times 10^1$	$1\cdot8 \times 10^2$	$3\cdot6 \times 10^2$	$7\cdot2 \times 10^2$
ϵ	1·1	1·8	2·2	2·2
t/s	$2\cdot2 \times 10^3$	$7\cdot2 \times 10^3$	$3\cdot6 \times 10^4$	$7\cdot2 \times 10^4$

11

Radioactivity, Radiation Detection and Dosimetry

An atom consists of a central, positively charged nucleus surrounded by negatively charged electrons. An atom becomes excited when electrons are raised to higher energy states. The excitation energy is lost by the emission of photons (electromagnetic radiation). The nucleus can also be excited and in this case the excitation energy is lost by the emission of particles (α- and β-radiation) as well as photons (γ-rays). The emission of radiation by an excited nucleus is called radioactive decay.

11.1 The structure of the atomic nucleus

The nucleus is made up of two types of particles, the proton and the neutron, which occupy a small region at the centre of the atom as shown in Fig. 11.1a. The proton carries a positive charge of 1.6×10^{-19} C (which is the same as the electronic charge but opposite in sign) and has a mass of 1.67×10^{-27} kg. The neutron is uncharged and slightly heavier than the proton. These nuclear particles are called nucleons. A nucleon is some 1840 times heavier than an atomic electron. The diameter of the nucleus is about 10^{-14} m, which is approximately 10 000 times smaller than that of the atom, so that the atom is mostly empty space. Since almost all the mass of the atom is in the nucleus, nuclear matter is extremely dense, of the order of 10^{17} kg m^{-3}.

A particular nucleus (or nuclide) is represented by the symbol $^A_Z X$, where X is the chemical element, A the mass number and Z the atomic number. The

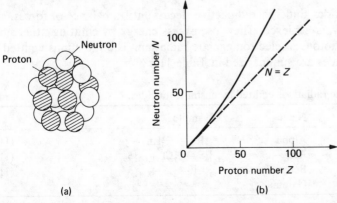

(a)   (b)

Figure 11.1 (a) A nucleus. (b) The variation of neutron and proton numbers for stable nuclides.

mass number is the total number of nucleons (protons plus neutrons) in the nucleus and Z is the number of protons. For example, $^{23}_{11}$Na represents a sodium nucleus containing 11 protons and 12 neutrons. All nuclei with the same number of protons belong to the same chemical element. However, nuclides of the same chemical element may have different numbers of neutrons. Such nuclides are called isotopes of that element. There are three isotopes of hydrogen, $^1_1$H, $^2_1$H (deuterium) and $^3_1$H (tritium), and the nuclides $^{10}_6$C, $^{11}_6$C, $^{12}_6$C, $^{13}_6$C, $^{14}_6$C and $^{15}_6$C are all isotopes of carbon.

The atomic weights of many elements are very nearly exact multiples of that of hydrogen. However, there are exceptions. For instance, the atomic weight of naturally occurring chlorine is $35 \cdot 5$, because this element comprises two isotopes, $^{35}_{17}$Cl and $^{37}_{17}$Cl, in the ratio 3:1. Similarly the atomic weight of neon is $20 \cdot 2$ because the isotopes $^{20}_{10}$Ne and $^{22}_{10}$Ne occur in the ratio 10:1 in the natural gas. The number of isotopes which are found in the natural form of an element varies considerably. Sodium and phosphorus have only one, hydrogen, carbon and nitrogen have two, and calcium has six.

Only about 15% of all known nuclides are stable, that is they remain unaltered with time. The two most important types of force operating within the nucleus are the short-range attractive force, which is called the strong nuclear force and exists between neighbouring nucleons and the long-range Coulomb repulsive force (§1.4) between the positively charged protons. Nuclear stability is governed by the balance between the attractive and repulsive forces and consequently depends upon the relative numbers of protons and neutrons in the nucleus. The variation of the neutron number $N = A - Z$ with proton number Z for the stable nuclides is shown in Fig. 11.1b. For light nuclei, stability is usual when the numbers of protons and neutrons are equal (e.g. $^{12}_6$C, $^{14}_7$N, $^{16}_8$O). In medium and heavy stable nuclei, the proportion of neutrons is increased (e.g. $^{75}_{33}$As, $^{127}_{53}$I, $^{238}_{92}$U) because the electrostatic repulsive force between the protons becomes more disruptive as Z increases and a larger number of neutrons is required to bind the nucleus and maintain stability.

11.2   Radioactivity and the decay law

Unstable nuclides undergo radioactive decay until a balance of forces, and hence stability, is achieved. They lose excess energy by emitting either small particles or photons of electromagnetic radiation. The radiations emitted by unstable nuclides are summarized in Table 11.1.

Table 11.1   The radiations emitted by unstable nuclides.

Symbol	Name	Example	
$^4_2\alpha$	Alpha	$^{226}_{88}$Ra $\rightarrow$ $^{222}_{86}$Rn $+$ $^4_2\alpha$	(11.1)
$^{0}_{-1}\beta^-$	Beta $-$	$^{35}_{16}$S $\rightarrow$ $^{35}_{17}$Cl $+$ $^{0}_{-1}\beta^-$	(11.2)
$^0_{+1}\beta^+$	Beta $+$	$^{33}_{17}$Cl $\rightarrow$ $^{33}_{16}$S $+$ $^0_{+1}\beta^+$	(11.3)
γ	Gamma		

Alpha-emission is common among the heavier nuclei which are too large to be stable. Excess energy is lost by emitting nucleons in the form of α-particles. These are $^4_2$He nuclei which consist of two protons and two neutrons. An example of this is the decay of $^{226}_{88}$Ra a nuclide which, if deposited in the body, can cause bone cancer. The disintegration can be described by Eq. 11.1 in Table 11.1 which shows that the total charge (the subscript) and the number of nucleons (the superscript) is unchanged in the transition. The radium nucleus is called the 'parent' nuclide and the radon nucleus is the 'daughter'.

Beta-emission covers three distinct processes. Nuclei which have too many neutrons for stability may lose energy by converting a neutron into a proton with the emission of a high-speed electron, a β^--particle, as described by the equation

$$^1_0n \rightarrow {}^1_1p + {}^{\,0}_{-1}\beta^-$$

Again note that there is no change in the total charge and the number of nucleons. This transition occurs in the decay of $^{35}_{16}$S to $^{35}_{17}$Cl as shown in Equ. 11.2. The parent nucleus, containing 16 protons and 19 neutrons, converts a neutron into a proton to produce a daughter with 17 protons and 18 neutrons.

Nuclei which have too many protons for stability may emit a positively charged electron, or positron (β^+), in transforming a proton into a neutron. The positron has all the physical properties of an electron except for its positive charge ($+1\cdot60 \times 10^{-19}$ C). The transition is written as

$$^1_1p \rightarrow {}^1_0n + {}^{\,0}_{+1}\beta^+$$

The formation of $^{33}_{16}$S from $^{33}_{17}$Cl is an example of this and is described by Equ. 11.3. As an alternative to positron-emission a nucleus may convert a proton into a neutron by capturing a K-shell electron. This process is called electron capture and $^{22}_{11}$Na may decay by this means:

$$^{22}_{11}Na + {}^{\,0}_{-1}e \rightarrow {}^{22}_{10}Ne$$

Gamma-emission is usual after a nucleus has altered its configuration by either α- or β-emission. The γ-ray is a photon of electromagnetic radiation of very short wavelength and is emitted when the daughter nucleus, formed in an excited state, falls to a lower energy level.

A radionuclide is usually produced by upsetting the proportion of neutrons and protons in a stable isotope of the element. The most common technique is to bombard the stable nuclei with neutrons which, being uncharged, may easily interact with a positively charged nucleus. A second advantage is their abundance in nuclear reactors. For example, when the nuclide $^{23}_{11}$Na captures a neutron, it is transformed into $^{24}_{11}$Na, the excess energy of the reaction being emitted as γ-radiation. The notation for this reaction is $^{23}_{11}$Na(n, γ)$^{24}_{11}$Na. Similarly the reaction in which $^{35}_{17}$Cl captures a neutron with subsequent proton emission is written $^{35}_{17}$Cl(n,p)$^{35}_{16}$S. Certain radionuclides may be produced by alternative reactions. The radioisotope of phosphorus $^{32}_{15}$P, used extensively in biological and clinical investigations, is obtained by either an (n, γ) reaction on $^{31}_{15}$P or by an (n, p) reaction on $^{32}_{16}$S, the method chosen depending on the reaction rate and on the required chemical form and purity of the product.

Once formed, a radioactive sample disintegrates at a rate peculiar to the radionuclide present in it. This disintegration rate $\Delta N/\Delta t$ (change in number of radioactive atoms per second) is found to be proportional to the number N of radioactive atoms in the sample at any time t, $\Delta N/\Delta t \propto N$. This can be written as

$$\Delta N/\Delta t = -\lambda N \tag{11.4}$$

where λ is the decay constant for the particular radionuclide and is the proportion of atoms decaying per second. λ is unchangeable by ordinary physical or chemical methods. The negative sign indicates that the population of radioactive atoms decreases with increasing time. If the number of radioactive atoms present at $t = 0$ is N_0 then the number N_1 present after a small interval of time Δt is N_0 minus the number of disintegrations that take place during this time. From Equ. 11.4 this is approximately $N_1 = N_0 + (\Delta N/\Delta t)\Delta t = N_0 - \lambda N_0\Delta t = N_0(1 - \lambda\Delta t)$ as shown in Fig. 11.2a. After another interval of time Δt the number $N_2 = N_1 + (\Delta N/\Delta t)\Delta t = N_1 - \lambda N_1\Delta t = N_1(1 - \lambda\Delta t) = N_0(1 - \lambda\Delta t)^2$. The number of atoms present after further intervals of time can be calculated in the same way and a disintegration curve drawn by joining the points $N_0, N_1, N_2 \ldots$. This is only approximate because we have assumed that during the time interval Δt the number N remains constant, equal to the value at the beginning of the time interval. This is not true. However if we make Δt very small the approximation is valid and a smooth curve is obtained as shown in Fig. 11.2b. This is called an exponential decay curve and the number of atoms N present at any time t is given by

$$N = N_0 e^{-\lambda t} \tag{11.5}$$

This equation can be derived mathematically, starting at Equ. 11.4 without making any approximations.

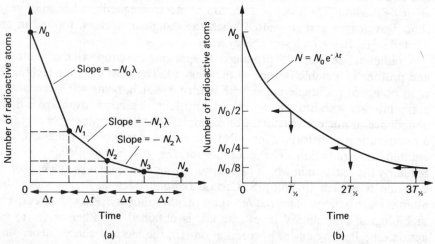

Figure 11.2　(a) An approximate radioactive decay curve. (b) The true decay curve.

The probability of decay is usually expressed in terms of the half-life $T_{1/2}$ rather than λ. This is the time required for half of any initial number of radioactive atoms to decay. Putting $N = N_0/2$ when $t = T_{1/2}$ in Equ. 11.5 gives $e^{\lambda T_{1/2}} = 2$ or

$$T_{1/2} = (1/\lambda) \log_e 2 = 0.693/\lambda \qquad (11.6)$$

After one half-life, one half of the original number of radioactive atoms remain, after two half-lives this fraction has fallen to one quarter, after three half-lives one eighth remain, and so on. Thus a radionuclide with a long half-life decays more slowly than one with a short half-life.

Half-lives vary from fractions of a second to many millions of years, but radionuclides with application in biology must necessarily have a more limited range. For example, the half-life of $^{131}_{53}I$ used to investigate thyroid function is 2·3 hours, that of $^{15}_{8}O$ used in respiratory studies is 2·1 minutes, whereas $^{14}_{6}C$, of importance in studying the metabolic behaviour of proteins, sugars and fats has a half-life of 5580 years.

11.3 The activity of a sample

The disintegration rate, or activity, of a radioactive sample is expressed in terms of the curie (Ci), which is that quantity of any radionuclide in which the number of disintegrations per second (dps) is 3.7×10^{10}. This unit (approximately the disintegration rate of 1 gram of radium) is to be replaced by the becquerel (Bq), where $1 \text{ Bq} = 1 \text{ s}^{-1} = 2.7 \times 10^{-11} \text{Ci}$. However, the curie is the current unit of activity and will be in common usage for some time. The curie is rather a large quantity and it is more usual to find activities expressed in submultiples of this unit. For instance, 1 mCi is an activity of 3.7×10^7 dps and 1 μCi an activity of 3.7×10^4 dps. Levels of activity applicable to medicine and biology are given in Table 11.2.

Table 11.2 Examples of levels of activity in various applications.

Activity	Application
10 nCi	Natural radioactivity in body
0·01–10 μCi	Samples from diagnostic tests
10 μCi–10 mCi	Patient scanning techniques
1–100 mCi	Cancer therapy with isotope implants
1000 Ci	γ-ray sources for radiotherapy
1 MCi	Sources for radiation processing

The activity A is numerically equal to the disintegration rate, $A = |(\Delta N/\Delta t)| = \lambda N$ so that, from Equ. 11.5, the activity at any time t is

$$A = A_0 e^{-\lambda t} \qquad (11.7)$$

where A_0 is the initial activity of the sample. The activity follows the same exponential decrease with time as the number of radioactive atoms present in the sample, and it is the quantity usually measured.

Example 11.1

The radioisotope of oxygen $^{15}_{8}O$ has a half-life $T_{1/2}$ of 2·1 minutes. (a) What is its decay constant? (b) How many radioactive atoms are present in a source of strength 4·0 mCi? (c) How much time will elapse before the activity in this sample is reduced by a factor of 8?

(a) From Equ. 11.6, $\lambda = 0.693/(2.1 \text{ min} \times 60 \text{ s min}^{-1}) = 5.5 \times 10^{-3}\text{s}^{-1}$.

(b) Since the activity $A = \lambda N$ the number of atoms $N = A/\lambda$ so that

$$N = \frac{4.0 \times 3.7 \times 10^{7}\text{s}^{-1}}{5.5 \times 10^{-3}\text{s}^{-1}} = 2.7 \times 10^{10}$$

(c) From Equ. 11.7, $1/8 = e^{-\lambda t}$ so that $e^{\lambda t} = 8$ or

$$t = \frac{\log_e 8}{\lambda} = \frac{2.079}{5.5 \times 10^{-3}\text{s}^{-1}} = 3.8 \times 10^{2}\text{s}$$

which is equal to three half-lives.

A radioactive sample might consist of the radionuclide alone (carrier-free), or it may also contain stable isotopes of the element in some chemical form. In the use of radioisotopes as tracers to study metabolism (literally to trace the path of a particular element in the body) the amount of chemical material to be administered must be kept to a minimum to maintain normal physiological conditions, and carrier-free material may be required. Radioactive iodine is frequently used to investigate thyroid function. Since the normal daily intake of iodine is about 150 μg, quantities much less than this must be used to maintain the physiological balance of the gland. The specific activity of a sample is $S = A/M$ where A is the activity and M the mass. The specific activity is a measure of the proportion of atoms in the sample which are radioactive and is usually expressed as $\mu Ci \, \mu g^{-1}$, $mCi \, g^{-1}$ or rather imprecisely for solutions as $mCi \, ml^{-1}$. The general rule for metabolic studies is that the smaller the animal, the higher should be the specific activity of the material used. In the study of vitamin-B$_{12}$ metabolism using radioactive $^{58}_{27}Co$, man has a daily intake of a few micrograms, and the specific activity of the tracer might be $1 \mu Ci \, \mu g^{-1}$. The daily intake of the rat is some 200 times lower, and the specific activity must be correspondingly higher if a similar monitoring system is used.

11.4 The interaction of ionizing radiation with matter

In atomic and nuclear physics the electronvolt (eV) is a commonly used unit of energy, where 1 eV = 1.60×10^{-19} J. Convenient multiples of this are keV(10^{3} eV) and MeV(10^{6} eV). The α-particles from a particular transition are monoenergetic, having a single, well defined kinetic energy in the range 2–9 MeV. γ-rays also have energies characteristic of the emitting nuclide, typically from several keV to a few MeV. However β-particles have no such well defined energies, the energies ranging from zero to a maximum value E_{max}, the average value of the energy of the β-particles emitted by a nuclide being about $(1/3) E_{max}$. The reason for this is that a particle called a neutrino (zero charge, near-zero mass) is emitted from the nucleus with the β-particle and the available energy is shared randomly between the two. Examples of maximum β-particle energies are 18 keV for tritium, $^3_1$H, and 1·7 MeV for $^{32}_{15}$P.

The atoms in a material are normally in the ground state, where the total energy of the orbital electrons is a minimum. An electron which gains energy may be excited to a higher energy level from which it will de-excite to the lower level with the emission of light or heat. If sufficient energy is available, the electron may be removed from the atom, a process called ionization (§ 3.1). This results in the formation of a free electron and a positive ion (an ion pair).

When a charged particle passes through a material there is an electrostatic interaction between the charge on the particle and the atomic electrons. The incident particle loses energy and is eventually absorbed by causing excitation and ionization of the atoms in the material. The particle loses about 35 eV of energy for every ionization produced, an average value which includes energy loss in the few excitations which occur between ionizations. In this way, a 3·5 MeV α-particle will produce some 10^5 ion pairs in losing all its energy and coming to rest at the end of its range. After coming to rest an α-particle picks up two surplus electrons and forms an atom of helium, whereas a β^--particle (really an electron) attaches itself to a positive ion. A positron, on the other hand, combines with an electron and both are annihilated, with the appearance of two γ-rays, each of energy 0·511 MeV (see Problem 11.10). An α-particle produces a large number of ion pairs over a very short range and may be absorbed by a few centimetres of air or a thin sheet of paper, whereas a β-particle produces fewer ion pairs over a relatively longer range. These ranges vary considerably due to the large differences in energies from different emitters. Thus the range in water of the most energetic β-particles from $^{32}_{15}$P($E_{max} = 1·7$ MeV) is 7 mm whereas the corresponding range for particles from tritium ($E_{max} = 18$ keV) is only 7×10^{-3} mm.

When electromagnetic radiation, in the form of X-ray or γ-ray photons, passes through matter excitation and ionization are again the mechanisms by which energy is lost. There are three different processes: photoelectric absorption, Compton scattering and pair production. Unlike charged particles, photons have no well defined range in a material. Some may be absorbed after a short distance whereas others may travel much further before interaction. The intensity of a photon beam passing through a material is, from Section 4.8,

$$I = I_0 e^{-\mu L} \tag{11.8}$$

where I_0 is the incident intensity, I the intensity at depth L in the absorber and μ is the linear absorption coefficient.

Example 11.2

The linear absorption coefficient of body tissue for 1 MeV γ-rays is $7·0$ m^{-1}. Calculate the thickness of tissue which reduces the incident intensity by a factor of 2.

Putting $I = I_0/2$ in Equ. 11.8 gives $e^{\mu L} = 2$ or $L = (\log_e 2)/\mu = 0·693/7·0$ m$^{-1} = 0·10$ m.

This is the 'half thickness' of tissue at this energy. This value may be compared with that for lead ($L_{1/2} = 7.9 \times 10^{-3}$m, $\mu = 88$ m^{-1}) which is about ten times less. This high absorption of photons by lead is the reason why it is used as a shielding material around radiation sources.

The range of ionizing radiation in a given material varies with the type of particle and its energy. Since the range of a β-particle in tissue is much greater than that of an α-particle of the same energy, the β-particle produces a much smaller number of ion pairs per unit length of its range. The energy lost per unit distance in the material is the linear energy transfer (LET) of the radiation. β-particles and γ-rays have relatively low LET values compared with α-particles, an important factor in dose assessment (§ 11.9).

11.5 Gaseous detectors

Quantities of radioactivity are measured by detecting the radiations emitted by the source or sample. The observed count rate (number of particles detected per second) may then be related to the activity of the sample. The following paragraphs describe methods of detecting β-particles and γ-rays, since α-emitters are rarely used in biological studies.

When ionizing radiation passes through a gas contained in a vessel as shown in Fig. 11.3, some of the atoms are ionized. In the absence of an electric field the positive ions recombine with the electrons. If an electric field is established between two electrodes placed in the gas, by applying a voltage E_b to them, the electrons move towards the positive plate and the positive ions towards the negative plate (see § 1.4). If the two electrodes are connected by an external circuit, as shown, a current I flows in this circuit. The magnitude of this current is a measure of the ionization produced in the gas and is proportional to the intensity of the incident radiation. This is the principle of operation of the ionization chamber. The current flow caused by a single particle is very small and this type of detector usually measures the total current from a large flux of particles, which may be about 10^{-12} A.

When the electric field strength between the electrodes is increased (by increasing the voltage E_b) the electrons produced in the gas may acquire sufficient energy to cause further ionization by collisions with other, unionized, gas molecules. This 'gas multiplication' of ions increases the number of ion pairs collected, and the resulting voltage pulse from the detector is proportional to the energy of the radiation absorbed in the gas. Detectors operating in this way are termed proportional counters. They are not widely used in biology because extremely stable voltage supplies are required to keep the gas multiplication constant. As the electric field strength between the electrodes is further increased, more and more electrons are released, each causing further ionization. The gas multiplication eventually reaches saturation and an 'avalanche' of electrons spreads along the length of the positive electrode. The output voltage is then independent of the energy deposited in the detector. This is the principle of operation of the Geiger–Müller counter.

A Geiger–Müller tube is shown schematically in Fig. 11.3. It consists of a metal tube with a central, wire, electrode and a thin mica window. The tube is filled with a gas at low pressure. In its normal state the gas is an electrical

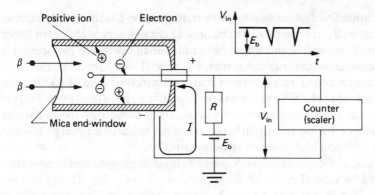

Figure 11.3 A schematic diagram of a Geiger–Müller (end-window) tube with counter.

insulator so that no current flows through the external circuit, $I = 0$, and there is no voltage drop across the resistor. When a particle enters the tube, ionization takes place and an avalanche of ions is produced which causes a current pulse to flow through R. This generates a voltage pulse across R. The voltage, V_{in}, across R is fed to an electronic counting circuit which counts the number of pulses that occur in a given time period. The variation of count rate with voltage E_b applied to the tube exhibits a plateau region where the count rate is independent of E_b as shown in Fig. 11.4a. The tube is operated in this plateau region. This reduces the necessity of high supply stability and, since the output pulses are large enough to be counted without additional amplification, this detector is used extensively, both in the laboratory and the field. The Geiger counter, filled with an inert gas, usually argon, is an ideal detector of β-radiation, registering about 99% of all particles entering the sensitive volume. For photons the detection efficiency is typically less than 1% because of the high penetrating power of these radiations.

The avalanche of electrons towards the positive electrode would normally continue for an appreciable time, but is usually terminated artifically by an external quench unit which reduces the voltage on the detector and prevents further gas multiplication. In this condition, the detector is unreceptive to

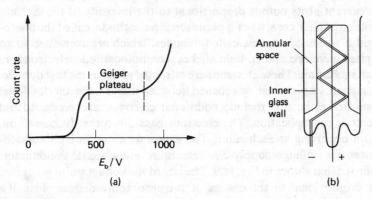

Figure 11.4 (a) Variation of count rate with tube voltage for a typical Geiger tube. (b) A liquid sample Geiger tube.

further radiation for T seconds after each pulse, where T is the 'paralysis time' of the detector. If, in a particular measurement, a count rate of N counts per second (cps) is recorded, the actual 'live time' of the detector per second is $(1 - NT)$ seconds, so the true count rate is $N_0 = N/(1 - NT)$ cps. For example, if 1000 cps are recorded with a detector having a paralysis time of 100 μs, then the true count rate is $1000/0.90 = 1100$ cps. Use of the above formula to correct observed count rates should rarely be required in measuring the radiations from samples of biological origin since it would indicate a poorly designed experiment with too high an activity being used.

Three types of Geiger tube are commonly used in biological measurement: the end-window tube (Fig. 11.3), the liquid sample tube (Fig. 11.4b), and the gas tube (which is not shown). In the end-window tube counter radiation enters through a window of mica sufficiently thin to allow low energy β-particles, such as from $^{14}_{6}C(E_{max} = 158$ keV) and $^{35}_{16}S(167$ keV), into the sensitive volume of the detector. This type of geometry is employed for solid samples, which are often contained in small planchettes placed close to the window. Liquids are placed in the annular space of a liquid sample counter. The detection efficiency is reduced by the absorption of β-particles in the inner glass wall. The advantage of this type of detector is that it avoids the necessity of having to reduce a liquid sample to solid form. If the radioactive sample is a gas, e.g. $^{14}_{6}CO_2$ or perhaps a tissue specimen which may readily be combusted to gaseous form, it may be incorporated into the filling of a gas counter. This provides a high detection efficiency.

11.6 Solid detectors

Gas-filled detectors are largely transparent to X-rays and γ-rays because of the high penetrating power of these radiations. A solid detector offers a higher probability of absorption and is consequently more sensitive to photons than a Geiger counter.

In the scintillation counter (Fig. 11.5a), the radiation excites atoms in a block of solid phosphor, a material in which excited atoms emit light when they de-excite. A crystal of NaI is a suitable phosphor for detecting γ-photons. The light emitted by the phosphor is directed into a photomultiplier tube which produces a current at its output proportional to the intensity of the incident light. A photomultiplier consists of a photosensitive cathode, called the photo-cathode, and a series of anodes, called dynodes, which are contained in an evacuated glass envelope. When light strikes the photocathode, electrons are emitted from its surface. These electrons are attracted towards the first dynode, because of a positive potential established between it and the cathode. When the electrons strike the first dynode additional electrons are produced and there is electron multiplication. The electrons pass down the dynode chain, multiplication occurring at each stage. The external circuit for a photomultiplier consists of a voltage supply E_b, resistor R and a signal conditioner/display similar to that shown in Fig. 11.3. The size of the voltage pulse generated across R is proportional to the energy of the absorbed radiation. Thus if a 0.5 MeV γ-ray produces an electrical pulse of 10 mV, the absorption of a 1 MeV photon will produce a pulse of 20 mV. The scintillation counter provides a valuable means of distinguishing a particular radionuclide by the energy of its radiations.

Scintillation crystals are usually of cylindrical geometry, but for samples such as small volumes of blood, a well crystal shown in Fig. 11.5b provides a more efficient detection system. If a large quantity of liquid such as a radioactive urine sample is available, this may be assayed using the annular cup geometry of Fig. 11.5c.

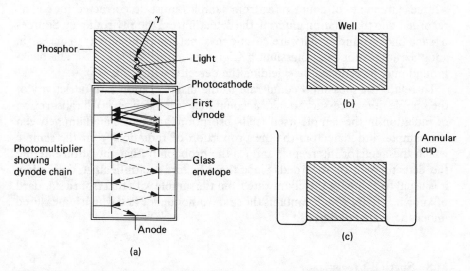

Figure 11.5 Scintillation detectors: (a) conventional cylindrical geometry showing electron pulse passing through photomultiplier; (b) well crystal; (c) crystal with annular cup.

11.7 Measurement of activity

The output from each of the detectors described is amplified where necessary and fed into a scaler, an instrument which records the number of voltage pulses over a period of time. In this way the count rate of the sample is obtained. The time at which any particular radioactive atom in the sample will disintegrate is uncertain, since radioactive decay is a random process and a measured count rate is a cumulative effect of many such disintegrations. Repeated measurements of the count rate of the same sample will differ. If the average count rate is 100 cps, repeated measurements might be 99, 102, 105, 101, 95, 98 . . . ; some might be as low as 80 or as high as 120. Because of this a single observation of a small number of disintegrations does not give a true indication of the activity of the sample. However, the statistical accuracy of the measurement increases as more counts are recorded. The spread of counts follows a normal distribution about the average count N, with a standard deviation $\sigma = \sqrt{N}$. Thus if $N = 100$, $\sigma = 10$ or $\sigma/N = 10\%$ whereas if $N = 10\,000$, $\sigma = 100$ or $\sigma/N = 1\%$. Since the percentage deviation decreases with increased counts, as many counts as possible should be recorded, although usually there is little point in accumulating more than 10 000 counts ($\sigma/N = 1\%$) because errors in pipetting and weighing then limit the accuracy achieved.

Example 11.3

The count rate from a radioactive urine sample is known to be roughly 5 cps. For how long must counts be recorded for the percentage deviation to be 2%?

The stipulation is that $\sigma/N = 2\%$ or $\sigma = \sqrt{N} = 0\cdot02N$. Thus $\sqrt{N} = 1/0\cdot02$ and $N = 2500$ counts. At the estimated count rate of 5 cps, this would take some 500 s, or a little over eight minutes.

All measurements of count rates from samples must be corrected for counts recorded due to radiation entering the detector from the environment. Sources of such background activity are cosmic rays, natural radioactivity in the earth, and the radioisotope of potassium $^{40}_{19}K$, present in the body (§ 12.5). This background may be reduced by shielding the detector with lead.

To relate the measured count rate to the disintegration rate or activity of the sample, it is necessary to take account of factors such as: (a) the absorption of radiation in the sample itself (self-absorption) and in the medium between the sample and detector; (b) the proportion of radiation from the sample which the detector intercepts; and (c) the probability that radiation entering the detector will be recorded. Since these factors are difficult to determine, it is usual to compare the count rate from the sample with that from a standard of known activity which contains the same radionuclide and which is measured under the same conditions.

11.8 Specialist techniques

The elements hydrogen, carbon and sulphur play a major role in biology, yet the only useful radioisotopes of these elements are the pure β^--emitters: $^3_1H(E_{max} = 18$ keV), $^{14}_6C(158$ keV) and $^{35}_{16}S(167$ keV) respectively. Only the highest energy β-particles from the carbon and sulphur isotopes can penetrate the window of a Geiger counter and all the particles from tritium are absorbed in the thinnest window. Since these nuclides emit no γ-rays a special detection technique, liquid scintillation counting, has been developed to enable these elements to be monitored. In this method the biological sample is mixed with a liquid phosphor. This reduces the problem of self-absorption in the sample and the detection efficiency is improved. The radioactive sample is dissolved in an organic solvent, usually toluene or xylene, along with an organic phosphor such as p-Terphenyl. The emitted β-particle is absorbed in the solvent, its energy is transferred to the phosphor and the pulse of light emitted is converted into an electrical signal using a photomultiplier tube, as described in Section 11.6. The radiations for which the technique is applicable have such low energies that relatively few electrons are emitted from the photocathode. Because electrical pulses arising from thermal emission of electrons in the photomultiplier tube may interfere with such small signals, this noise can be reduced by cooling the tube.

A second technique is one which determines the position of radioactive substances within biological specimens. This is autoradiography, where a radioactive histological specimen is positioned close to a photographic emulsion. The energy of the emitted radiation transforms the silver bromide to photolytic silver which is developed into black silver grains. The advantages of this technique are that very small quantities of radioactivity may be detected if the exposure is sufficiently long, and the resultant photograph is a permanent two-dimensional record of the measurement. Autoradiography is ideal for localizing low energy β-emitters in tissue since the ranges of the particles in the emulsion are small, resulting in a high resolution of the activity.

If an autoradiograph of animal tissue is required, the radionuclide (commonly tritium, $^{14}_{6}C$, $^{35}_{16}S$ or $^{125}_{53}I$) may be administered orally or by injection. The animal is then frozen and the required thin sections cut with a microtome. For plants, nuclides commonly used are $^{14}_{6}C$, $^{32}_{15}P$, $^{35}_{16}S$ and $^{45}_{20}Ca$. The plant is then either frozen and sections cut, or the whole plant is placed in contact with the emulsion.

Although quantitative determinations of activity may be made either by densitometry, where the optical density of the film is measured using a photometer, or by counting the number of developed silver grains in some small area of the film, a popular exercise is to relate the autoradiograph to the histological detail of the specimen. For example, if the tissue section is stained and the emulsion developed photographically, the result of the combination when viewed under an optical microscope shows darkened areas of exposed film superimposed on the tissue specimen. If the sample and emulsion are suitably prepared for use on an electron microscope, an electron micrograph shows the localization of activity in the tissue in correspondingly greater detail.

11.9 Dose assessment

Ionizing radiation interacts with matter by producing ion pairs and causing excitations. When the body is exposed to radiation, energy is deposited in tissue, one ionization corresponding to an energy deposition of about 35 eV. The unit of absorbed dose in current use is the rad which corresponds to an energy absorption of 1.0×10^{-2} J in each kilogram of tissue. This will be replaced by the gray (Gy) which is the absorbed dose when 1.0 J is deposited per kilogram. Hence 1 Gy = 100 rad.

Example 11.4

Estimate the number of ionizations produced per kilogram of tissue for an absorbed dose of 1.0 rad.

1.0 rad corresponds to an energy deposition of 1.0×10^{-2} J kg^{-1}, and since about 35 eV is required per ionization, the number of ionizations is

$$N = \frac{1.0 \times 10^{-2} \text{J kg}^{-1}}{35 \text{ eV} \times 1.6 \times 10^{-19} \text{J eV}^{-1}}$$

$$N = 1.8 \times 10^{15} \text{kg}^{-1}$$

Such a simple unit of dose as the rad is quite satisfactory for photon and β-particle dosimetry. However, an important factor in the biological damage produced is the rate of ionization along the track of a particle. This is the linear energy transfer LET ($\S11.4$). Because α-particles have a higher LET than the other radiations the biological damage produced by an α-particle is higher than for a photon of the same energy. The relative biological effectiveness (RBE) is a measure of the relative damage caused by a particular radiation compared with that caused by the same absorbed dose of photons. It is defined as

$$\text{RBE of radiation} = \frac{\text{dose of photons required to produce a given effect (rad)}}{\text{dose of radiation required to produce same effect (rad)}}$$

In general the RBE depends on the damage considered, the LET and the energy of the radiation. Usual working values are 1 for photons and high energy β-particles, 2 for low energy β-particles, 10 for fast neutrons and protons and in the range 10–20 for α-particles.

The biological damage produced is measured in terms of the effective dose, the rem, where

$$\text{effective dose (rem)} = \text{absorbed dose (rad)} \times \text{RBE}$$

For example, an absorbed dose of 1 rad of fast neutrons (RBE = 10) corresponds to an effective dose of 10 rem and produces ten times the damage of the same absorbed dose of γ-rays. Although the same amount of energy is deposited in each case, the effect of the neutrons is greater because the ionization is not produced uniformly but concentrated in dense tracks which will cause more serious changes in tissue.

Example 11.5

Neutron activation analysis is a technique which may be applied to determine the amounts of several elements in the body. When tissue is irradiated with neutrons, some stable atoms become radioactive and the number of γ-rays subsequently detected is a measure of the quantity of an element in the body. If neutrons of average energy 2·5 MeV and RBE of 10 are used in the procedure, estimate the number of neutrons absorbed per kilogram of tissue if an effective dose of 0·50 rem is to be given to the patient. Assume all neutrons are absorbed in the body.

Since effective dose = absorbed dose × RBE, the absorbed dose = 0·50/10 = $5·0 \times 10^{-2}$ rad, which corresponds to an energy deposition of $5·0 \times 10^{-4}$ J kg^{-1}. Each neutron has energy $E_n = 2·5 \times 10^6$ eV and since 1 eV = $1·6 \times 10^{-19}$ J, $E_n = 2·5 \times 10^6$ eV × $1·6 \times 10^{-19}$ J eV^{-1} = $4·0 \times 10^{-13}$ J. Therefore, the number of neutrons required is

$$N = \frac{5·0 \times 10^{-4} \text{J kg}^{-1}}{4·0 \times 10^{-13} \text{J}} = 1·3 \times 10^9 \text{kg}^{-1}$$

The assumption that all neutrons are completely absorbed is unrealistic. Most neutrons will lose only part of their energy before being scattered out of the body and some will pass through without interaction.

The radiation dose absorbed by the body is measured by a dosemeter. This is basically a detector whose response is modified to give the radiation dose directly in rad or rem. Some instruments provide the dose rate, usually in mrem h^{-1} or μrem h^{-1}. An ideal dosemeter should have a response which is proportional to dose over a wide dose range and is independent of the type of radiation and its energy. In practice the instrument chosen will depend on the radiation and its energy and probably on the ease of use and portability of the dosemeter. A common dosemeter is the personal film badge used by people routinely exposed to radiation, the blackening of a photographic film being related to the dose received. Such dosemeters measure not only the total dose received over a specified period, but also give some indication of the contributions of different types of radiation so that a more realistic effective dose may be obtained.

Problems

11.1   Determine the atomic weight of copper from its natural composition: 70% $^{63}_{29}$Cu and 30% $^{65}_{29}$Cu.

11.2   If $\Delta N/\Delta t = -\lambda N$, show that $N_m = N_0(1 - \lambda \Delta t)^m$ after m intervals of time Δt. Noting the total elapsed time is $t = m\Delta t$ and that by definition e$^x = (1 + x/m)^m$ for large m, show that $N = N_0 e^{-\lambda t}$.

11.3   A sample of $^{128}_{53}$I contains $2 \cdot 0 \times 10^{10}$ radioactive atoms. If the half-life of this isotope is 25 minutes, calculate the number of atoms decaying per second.

11.4   The volume of extracellular fluid may be measured by injecting $^{35}_{16}$S-labelled sodium sulphate. One such source has an initial activity of 2·0 mCi. If this isotope has a half-life of 87 days, calculate the activity of the source after 60 days in curies and in becquerels. What time elapses before the activity falls to 0·50 mCi?

11.5   The decrease in activity of a sample with time is $A = A_0 e^{-\lambda t}$. Show that a graph of $\log_{10}A$ against t is a straight line of gradient $-\lambda/\log_e(10)$. A Geiger counter detects positrons from the decay of a sample of $^{13}_{7}$N, the decrease in activity with time being given below. Plot a graph of $\log_{10}A$ against t to find the decay constant and half-life for this nuclide.

Activity/dps	1000	880	750	650	590	510	430	370	340
Time/minutes	0	2	4	6	8	10	12	14	16

11.6   Living wood utilizes carbon in its metabolic cycle of which the isotope $^{14}_{6}$C (produced by cosmic rays in the atmosphere) is $1 \cdot 35 \times 10^{-10}$% abundant. If the half-life of this nuclide is $5 \cdot 58 \times 10^3$ yr, calculate its decay constant (per minute). The atomic weight of carbon is 12·01. Determine the number of disintegrations per minute in 1·00 kg of carbon. ($N_A = 6 \cdot 022 \times 10^{26}$ atoms kmol^{-1}.)

11.7 Radiocarbon dating is based on the assumption that the ratio of $^{14}_{6}C$ to stable carbon atoms in the atmosphere has remained constant for thousands of years, and that all living organisms contain this prescribed ratio of carbon atoms. Upon death, $^{14}_{6}C$ incorporation ceases and the amount then present decays at the characteristic rate. Therefore the determination of the specific activity of carbon fixes the time of death. The specific activity of a sample of carbon from a wood relic is 4·5 nCi kg^{-1}. Use the result of Problem 11.6 to determine the age of the relic.

11.8 What proportion of the incident intensity of a beam of 1 MeV photons is transmitted through 0·20 m of body tissue? (Refer to Example 11.2).

11.9 Estimate the number of ion pairs produced in the gas filling of an ionization chamber by: (a) an 8·0 MeV α-particle; and (b) a 150 keV β-particle. Assuming no recombination of the ions, how much negative charge is collected on the positive plate in each case?

11.10 Under certain circumstances mass M and energy E are interchangeable, the relation being expressed in Einstein's equation $E = Mc^2$, where c is the speed of light. How much energy (a) in joules and (b) in eV is released when 1 kg of matter is converted into energy? Show that the energy equivalent to the mass of an electron or positron ($M = 9·11 \times 10^{-31}$ kg) is 0·511 MeV.

11.11 Explain why a Geiger counter is not useful for determining the energies of charged particles interacting in the filling gas.

11.12 A dried blood sample containing $^{32}_{15}P$ is assayed using an end-window Geiger counter which has a paralysis time of 100 μs. If the average count rate is 50·0 cps and the sample is counted for 2·0 minutes find: (a) the true total counts expressed as a percentage of the observed total counts; and (b) the percentage accuracy of the measurement. Comment on the results.

11.13 A standard source containing 2·0 μCi of tritium is monitored by a liquid scintillation counter and 120 cps are recorded. Calculate the activities of samples of the same chemical form which give count rates of 150, 270, 30 and 540 cps.

11.14 If a radioactive compound of high specific activity is mixed with a large quantity of the stable compound, the activity is diluted and comparison of initial and final specific activities enables the amount of the diluting compound to be estimated. This is the principle of isotope dilution analysis. In one procedure to measure total body water, 1·0 ml of tritiated water of activity 5·0 μCi was injected into the body. After an adequate time lapse to ensure thorough mixing, a 10 ml sample was withdrawn from the body and found to have an activity of 1·0 nCi. Estimate the mass of water in the body if the density of water is 1·0 $\times$ 10^3 kg m^{-3}.

11.15 It is found that the same degree of skin-reddening is produced by 13 rad of α-particles as by 200 rad of X-rays. Express these absorbed doses in grays. What is the RBE of α-particles for this damage? Calculate the effective dose of the α-particles in this case.

12

Radiobiology and Radiation Protection

When biological tissue is exposed to radiation, the energy absorbed may produce chemical changes in the cells which affect cell metabolism. These changes may be sufficiently damaging to lead to the death of individual cells and quickly produce a state of radiation sickness in multicellular organisms which may lead to death in extreme cases. Alternatively, there may be some permanent modification in cellular activity which could lead to genetic changes or to cancer induction years after exposure. For this reason the amount of radiation damage a person may receive from deliberate exposure is strictly controlled.

12.1 Effects of radiation on biological molecules

The initial damage caused by the interaction of radiation with tissue is at the molecular level. The formation of ionized molecules by the radiation leads to chemical changes in certain cell constituents. A mammal contains about 85% water by weight, therefore more of the energy will be deposited in water molecules than in the more complex molecules of the cell. These activated water molecules may produce chemical changes in larger molecules such as enzymes, proteins, nucleic acids and polysaccharides, which have greater biological importance. This mode of radiation damage is indirect action, in contrast to the direct action of the radiation on the large molecules themselves.

The first step in the activation of a water molecule is the primary ionization $H_2O \rightarrow H_2O^+ + e^-$. The electron is then captured by a neutral water molecule $H_2O + e^- \rightarrow H_2O^-$, the two neutral molecules being converted into H_2O^+ and H_2O^-. Each ion breaks down as shown below

$$H_2O^+ \rightarrow H^+ + OH^{\cdot}$$
$$H_2O^- \rightarrow H^{\cdot} + OH^-$$

The H^+ and OH^- ions recombine to form a neutral water molecule, leaving the net reaction

$$H_2O \rightarrow H^{\cdot} + OH^{\cdot}$$

$H^{\cdot}$ and $OH^{\cdot}$ are highly reactive free radicals, each trying to pair the single electron in the outer shell with that in another radical. Such free radicals have a very short lifetime (about 10^{-5} s in water), and may react in several ways. One possibility is recombination to form water $H^{\cdot} + OH^{\cdot} \rightarrow H_2O$, or the formation of a hydrogen molecule $H^{\cdot} + H^{\cdot} \rightarrow H_2$. Alternatively two $OH^{\cdot}$ radicals may combine to give hydrogen peroxide:

$$OH^{\cdot} + OH^{\cdot} \rightarrow H_2O_2$$

which is a stable molecule but an oxidizing agent which may cause damage if produced in a sensitive part of the cell, such as a chromosome in the cell nucleus. If a free radical is in the vicinity of an important organic molecule R—H, an organic free radical may be produced

$$R - H + OH^{\cdot} \rightarrow R^{\cdot} + H_2O$$

R˙ may be highly reactive and combine with other molecules.

The direct mode of radiation damage, where the energy is deposited in the organic molecule itself is more efficient in producing biologically significant changes. The radiation may cause a large polymer molecule (§10.2) to break into smaller fragments (main-chain scission). Frequently the break occurs in the weakest bond in the molecular chain, indicating that the deposited energy may be transferred along the molecule to this bond. If the radiation breaks a weak hydrogen bond

$$R - H \rightarrow R^{\cdot} + H^+ + e^-$$

the organic radical R˙ is produced which may form additional chemical bonds or crosslinks (see §10.2). At low molecular concentrations, the crosslinking is largely intramolecular as shown in Fig. 12.1a, the effective molecular contraction reducing the viscosity of the medium. At higher concentrations two R˙ radicals may be formed close together, and dimerize by forming an intermolecular crosslink.

$$R^{\cdot} + R^{\cdot} \rightarrow R - R$$

with consequent increase in viscosity. This is shown in Fig. 12.1b. As the radiation exposure or molecular concentration further increases, the viscosity of the medium increases until a gel is formed.

(a)   (b)

Figure 12.1   (a) Intramolecular and (b) intermolecular crosslinks in polymers.

If oxygen is present near the site of the initial chemical change, the probability of damage is increased. This may be due either to an increase in the amount of hydrogen peroxide produced

$$O_2 + H^{\cdot} + H^{\cdot} \rightarrow H_2O_2$$

or by a direct chemical reaction with the organic free radical

$$R^{\cdot} + O_2 \rightarrow RO_2^{\cdot}$$

where the formation of the peroxide radical $RO_2^{\cdot}$ may cause more severe damage than R˙ alone.

Oxygen is a chemical sensitizer, in that the effects of the radiation may be worsened by its presence. There are several substances which lessen those effects, an example of such a chemical protector being cysteamine which contains an —SH group. A possible mechanism for its action is

$$R^{\cdot} + P - SH \rightarrow R - H + P - S^{\cdot}$$

where the damage sustained by the important organic molecule is transferred to the protector.

12.2 Effects of radiation on cells

The cell is the basic unit of life. Since the effects of radiation on a multicellular organism are mainly due to damage to its component cells, an understanding of the effects of radiation at cellular level is important. These effects are summarized in Table 12.1.

Table 12.1 Radiation effects on the cell.

Cell death	Growth and function modification
Mitotic death	Mitotic delay
Interphase death	Increase in membrane permeability
	Gene mutation
	Chromosome breakage

When an animal grows, the size of each individual cell does not alter greatly. Growth is achieved by an increase in cell number. Mitosis is the process of cell division which occurs in the somatic (non-genetic) cells in which both nucleus and cytoplasm of one cell divide to form two new cells. The cell nucleus contains chromosomes carrying the genes of the individual, and before mitosis each chromosome is duplicated to form two chromatids, one chromatid going to each daughter cell. In this way all the genetic information is transferred to the next generation. The various cytoplasmic structures also duplicate and separate. After mitosis, each cell goes through a period of growth (interphase) before the next mitosis. The periods of interphase and mitosis constitute the cell cycle. Cell multiplication by mitosis continues until an organ is the required size, but in organs such as the bone marrow, gastro-intestinal tract and skin, additional cell division is necessary to replace damaged cells. The stem cells in such organs undergo mitosis regularly to replenish the cell population. Certain cells such as nerve and muscle cells become highly differentiated and perform specialized functions. Such cells no longer divide spontaneously. In general the sensitivity of cells to radiation is directly proportional to their reproductive activity and inversely proportional to the degree of differentiation. For example, the mature, specialized cells on a villus of the intestinal epithlium are more radioresistant than the rapidly dividing stem cells from which they originate.

Cell death is the most obvious effect of radiation. Two types of cell death may be induced: mitotic death and interphase death. After irradiation the cell may undergo mitosis once or twice but then loses its ability to reproduce, and is sterile. This is mitotic death. The cell still metabolizes, and may grow to several times the normal cell size. Mitosis may be regarded as the weak link in the cell cycle at which any damage incurred during irradiation may become apparent. The variation of mitotic death rate of a pure cell line with dose is shown in Fig. 12.2a. The apparent resistance of some of the cells is due purely to chance, since they are all identical. Only a certain number of 'hits' can be made from a given amount of radiation, the cells surviving being the ones not 'hit'. Therefore it is unrealistic to quote the dose required to kill all cells in a population, since usually a few will escape any amount of radiation. For bacterial spores, the radiation dose required for 90% of the cells to suffer mitotic death may be as high as 500 000 rem, whereas for the same spores in water, this dose falls to about 100000 rem, showing the contribution of indirect action in this case. The dose for mammalian cells grown in culture varies from 100 to 1000 rem, depending on cell type. These figures are for cells irradiated outside the body, and cannot be taken as representative of radiosensitivities *in vivo* where cell metabolism depends on the secretions of other cells.

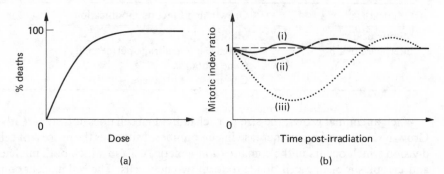

Figure 12.2 (a) Dependence of cell death on radiation dose. (b) Mitotic delay. Dose increases from (i) to (iii).

In interphase death the cell is prevented from entering mitosis at all, and no division occurs after irradiation. Highly differentiated cells which do not divide can only be killed by interphase death since mitotic death cannot occur. For this reason nerve and muscle cells are radioresistant, doses of several thousand rem being required for death. However the lymphocyte, which does not divide, is very radiosensitive. The closely related amoeba, with a high reproductive activity is far more resistant. Besides differentiation and reproduction, it is evident that other properties such as number of chromosomes or number of mitochondria in the cytoplasm are likely to influence the probability of cell death.

Radiation may induce changes in cell cycle and structure at doses insufficient to kill. Relatively low doses may result in mitotic delay, a temporary change in normal cell cycle. In a population of identical cells the fraction undergoing mitosis at any time is the mitotic index which is normally constant if temperature and nutritional environment are unaltered. If a population of such cells is irradiated and its mitotic index compared with that of a control population, the ratio is the mitotic index ratio (MIR).

The change in MIR with time after irradiation is shown in Fig. 12.2b for several radiation doses. The drop in MIR is more pronounced and of longer duration for higher doses. This reduction is caused by radiation delaying the onset of mitosis in some cells. At higher doses more cells are delayed. After some time there is a compensating wave as delayed cells begin to divide along with those whose cycle was unaffected.

Gene damage is a second modification caused by radiation. The genetic information of the cell is carried by the genes located on specific loci of the chromosomes. Radiation may alter the chemistry of a gene and cause a mutation. Such mutant genes are unobservable directly, the effects only becoming apparent after a long time. The majority of mutations are harmful and possibly lethal, though occasionally a desirable mutation may be induced artificially as in the development of a strain of barley with a high resistance to rust disease. Although mutations occur naturally and this process is a basic mechanism of evolution, the effect of radiation on mutation rate is significant.

A second form of genetic damage occurs when the radiation causes chromosome 'breakage'. The cell is most sensitive to this damage when irradiated during interphase when the chromosomes are not visible. The radiation induces a chemical change which is amplified by metabolism, resulting in the observed anatomical scission. The broken fragment may either rejoin the original chromosome, so repairing the damage, or form a new structure with a different chromosome. This may lead to complications at the anaphase stage of mitosis when the chromatids separate. Alternatively the fragment may remain isolated and form a micronucleus in one of the daughter cells. The severity of the damage depends on the genetic information lost at mitosis. The worst situation is when such chromosome abberations occur in the genetic stem cells. Few problems would result in the somatic blood stem cells in the bone marrow if the information lost related only to eye colour, for example.

In all cases radiation induces some primary chemical lesion which affects the metabolism of the cell, the damage becoming observable at some later time. Since ionizing radiation is non-selective, every molecule in the cell has a chance of sustaining damage. Since the doses required to produce most changes are sufficient to affect relatively few molecules, it is likely that damage to the larger, essential molecules constitutes the primary lesion. Such essential molecules include the nuclear DNA and certain key enzymes regulating the energy supplies of the cell. Alternatively radiation may rupture important intracellular barriers. The amount of energy required for this indicates why high LET radiations have a high biological effectiveness (§ 11.9). Lysosomes in the cytoplasm may be regarded as small packets containing enzymes able to hydrolyse some proteins and nucleic acids. Such enzymes are normally confined to the lysosome and cause no damage. Lysosome walls have been found to be ruptured in cases of cell death, and this enzyme-release theory explains many observed radiation effects, especially interphase death.

If cells are irradiated continuously at low dose rates, a higher total dose is required for death than if the radiation is given in a single dose. The cell has some capacity to repair certain lesions provided the dose rate is not excessive. There are essentially two types of damage. The first may be repaired if the cell has sufficient time but the second, such as a gene mutation, is irreparable and therefore cumulative.

12.3 Radiation sickness and death in animals

Radiation effects on multicellular organisms may be divided into two categories: acute effects which are apparent shortly after irradiation and chronic or long-term effects which may arise many years later. These effects are summarized in Table 12.2. There have been relatively few documented cases of severe acute effects in man e.g. the results of atomic-bomb explosions. The only data available for which doses are accurately known have been obtained from studies of radiation effects in other mammals, mainly mice and rats. However it has been shown that such effects are similar to those in man.

Table 12.2   Radiation effects on multicellular organisms.

	Effect	Cause
	Anaemia, haemorrhages	Reduction in erythrocytes and platelets (a)
	Susceptibility to infection, weight loss	Reduction in white blood cells and cells lining intestine, skin damage (b)
ACUTE	Sterility	Mitotic death of spermatogonia and oocytes
	Death	(a) or (b) above or destruction of CNS
	Genetic mutation	Chemical lesion in nucleic acid
	Embryo damage	Irradiation at organogenesis
CHRONIC	Cancer induction	Somatic mutations or favourable environment
	Cataract formation	Radiation causes clouding of eye lens

Although it is difficult to extrapolate from cellular to whole-body effects because of the interdependence of cells in the body, organs containing cells with high mitotic activity are the most radiosensitive and it is in such organs that the effects of radiation first become apparent. For example, the skin is sensitive to low doses of radiation, a localized dose to this organ resulting in reddening very similar to sunburn. At high doses this may lead to sores and blistering, hair loss and pigment de-activation, with ulceration and cancer induced later.

The bone marrow is a radiosensitive organ containing stem cells which divide rapidly to maintain cell levels in the blood. Fig. 12.3a shows the reduction in blood cell population in a mammal following whole-body irradiation. Lymphocytes are very sensitive to interphase death and their reduction is almost immediate. Granulocytes and erythrocytes are relatively resistant but mitotic death in their stem cells causes the later drop in population. The acute reduction in white blood cells leaves the body susceptible to infection, and the low erythrocyte number, together with a corresponding platelet reduction, results in reduced oxygen transport, haemorrhages and anaemia.

Although the intestinal cells which absorb nutrient are highly differentiated and so relatively resistant, their stem cells are radiosensitive. When the cells of the intestine are worn away, radiation restricts their replacement and bacteria from the gastro-intestinal tract enter the body through the intestine walls, so leading to infection. The chances of infection are greatly increased by the

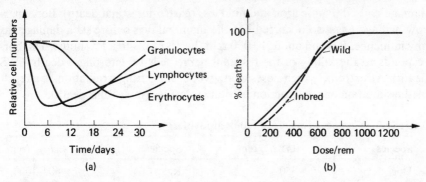

Figure 12.3 (a) Post-irradiation blood cell reduction in a mammal. (b) Variation of death rate of mice with dose.

reduction in white blood cells. Radiation causes the intestine to become leaky and body fluids may be lost through the walls, resulting in weight loss.

A non-lethal effect of radiation is sterility. The male may remain fertile for several weeks after irradiation since the mature sperms are relatively radioresistant, but as the stem cells are susceptible to mitotic death there may then be a period of sterility lasting several months. A high localized dose of radiation to the testes alone may produce permanent sterility. The ovaries are more radiosensitive than the testes.

The dose required to kill a multicellular organism is very variable. Animals may be classified as radiosensitive or radioresistant. Radiosensitive animals rely on cell division in some organs to maintain life. All vertebrates fall into this category, the bone marrow and intestine being the sensitive organs in this respect. Cell division does not occur in the adult stage of some animals, notably the adult insect in which all somatic cells are in a post-mitotic state. Such cells can only be killed by interphase death requiring large doses, and these animals are resistant. Fig. 12.3b shows the variation in mortality of mice with dose for a single irradiation. No deaths occur at low doses but the mortality progressively increases to 100% at high doses. Two curves are shown, one for a population of wild mice and one for a closely-inbred strain. This illustrates the dependence of mortality on genetic factors. Unlike a population of identical cells, the animals in a species have a range of radiosensitivities because of natural variation. Also they may die at different times after irradiation. Therefore the radiosensitivity of a species is usually expressed as the $LD_{50/30}$, the single lethal dose of whole body radiation required to kill 50% of the animals within 30 days. The significance of this time is that few animals will die after about three weeks, and such deaths would not be considered as acute effects. Values of $LD_{50/30}$ for several species are shown in Table 12.3.

Both the probability of an animal dying from a single radiation exposure and the time of death depend on the dose received. Fig. 12.4 shows the mean (average) survival time (MST) plotted against dose for a typical mammal. At a dose of about 100 rem, the MST drops sharply to a plateau at roughly 15 days. In this dose region the mammal dies from bone marrow damage (haemopoietic death). The MST drops further with increasing dose until at about 1000 rem a second plateau is reached corresponding to death at three to four days due to intestinal damage caused by infection and fluid less (gastro-intestinal death). Bone marrow damage has also occurred but the mammal dies before such damage has any influence. Beyond about 10 000 rem the MST falls to a few hours. This corresponds to the killing of the resistant nerve cells by interphase death (CNS death). At extremely high doses death will occur during irradiation as a result of the simultaneous destruction of vital molecules (molecular death).

Table 12.3   Comparative animal radiosensitivities.

Species	$LD_{50/30}$/rem	Species	$LD_{50/30}$/rem
Dog	350	Rat	600–1000
Mouse	400–600	Frog	700
Monkey	600	Newt	3000
Man	600–700	Snail	8000–20000

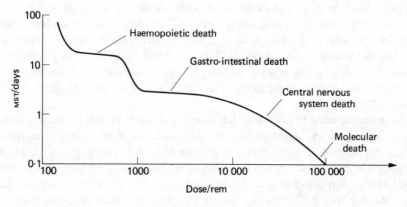

Figure 12.4   Variation of mean survival time with dose for a typical mammal.

The condition of radiation sickness in an animal caused by stem cell death in the sensitive organs may lead to death of the animal unless the damage is repaired or its effects modified. The viable stem cells remaining after irradiation divide and attempt to restore the numbers of mature cells to former levels, the degree of success depending on the harm caused by such reductions in the meantime. Radiation damage can be reduced by a chemical protector, but such a protector must be present in the cell at the time of the primary lesion. The risk and extent of infection due to intestine and skin damage can be reduced by treatment with antibiotics, and blood loss and anaemia by platelet transfusion to restore blood-clotting. A bone-marrow transplant may be possible in subjects with depleted blood stem cells, the aim being to repopulate the bone marrow with new cells and alleviate damage. However, such 'grafts' will only be successful if radiation has destroyed the immune response of the host, otherwise the stress imposed from the transplant of foreign material can have a more serious effect than the radiation itself, as the immune response recovers and rejects the transplant.

12.4   Long-term effects of radiation

The previous section has described some radiation effects which occur relatively quickly. An individual may recover from such radiation sickness and suffer long-term injuries which become apparent many years later.

One such chronic damage is the genetic mutation discussed in Section 12.2. The probability of a mutation is proportional to the total, cumulative dose received. The cell is unable to repair a mutation as such, although some mutations may prevent the cell proliferating and so restrict the damage. A second effect which, although not necessarily long-term, may be regarded as damaging to the next generation occurs when the embryo is irradiated. The embryo of any species is generally more radiosensitive than other stages of development since the component cells exhibit high mitotic activity. The main damage results from irradiation at organogenesis (2–7 weeks post-conception in man), when the organs of the embryo are formed, with a high probability of severe abnormalties at birth.

Another important chronic effect is cancer induction. A cancer or tumour cell is not under the control of the body, and is produced in numbers surplus to requirements. This implies a high division rate, although many tumours have a lower mitotic activity than normal cells in bone marrow or intestine wall. It is essentially the lack of control which distinguishes tumour cells from normal cells. Certain tumours such as cysts and warts are benign and normally harmless, but cancers are malignant tumours, the cells of which have the ability to break from the main growth and form secondary growths or metastases in other parts of the body. There are several possible mechanisms of cancer induction by radiation. For instance, tumours may result from somatic mutations leading to uncontrolled mitosis, or radiation may create a favourable environment for malignant cells, of whatever origin.

Although there is always a long latent period between irradiation and cancer growth, the effects of localized and whole-body irradiations are quite different. For a localized irradiation of one part of the body there appears to be a certain threshold dose (usually in excess of 1000 rem) below which a cancer is not induced. The chronic damage occurs in the irradiated organ, a bone tumour resulting from skeletal irradiation or cancer of the thyroid from localized irradiation of that gland. The type of cancer induced following whole-body irradiation depends largely on genetic factors. For example, leukaemia (an excess of white blood cells), ovarian tumours and mammary gland tumours are prevalent in different strains of mice.

Besides being an agent of cancer induction, ionizing radiation is also used in cancer therapy, or treatment. Malignant cells, having a high proliferation rate, are sensitive to mitotic death and may be killed by this means. A common technique for therapy of tumours located deep in the body uses an external, multicurie source of γ-radiation which may be directed at the tumour. The damage to the surrounding normal cells may be reduced (tissue sparing) by fractionating the total dose given. This allows some repair of damage sustained by the surrounding tissue between fractions. Tissue sparing is further improved by using an irradiation geometry similar to that shown in Fig. 12.5, where succes-

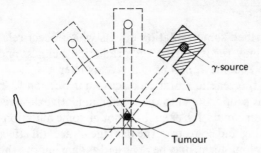

Figure 12.5 Rotation of radiation therapy source for tissue sparing.

sive fractional irradiations in different directions ensure that normal tissue receives less dose than the tumour.

12.5 Maximum permissible doses

Over the years there has been a dramatic increase in man-made sources of radiation. Some sources, such as nuclear reactors and sources for cancer therapy and medical diagnosis, are obviously beneficial. Since any exposure to radiation may have adverse effects, it is necessary to weigh these benefits against possible hazards. The hazardous effects of radiation may be acute, long-term or even passed to future generations. This latter genetic effect is the main consideration is assessing permissible dose levels to the population as a whole.

Besides exposure to artificial sources, man has always been exposed to radiation from the natural environment including cosmic rays, radium, thorium and uranium in the surrounding rocks and radon gas in the air. Additionally, the radioisotope of potassium, $^{40}_{19}K$, has a natural abundance of $0 \cdot 01\%$, and the body contains about $0 \cdot 14$ kg of elemental potassium. The average total dose to the gonads (the genetically significant dose) due to natural background is about 100 mrem per year but, in parts of the world were the rock contains a lot of radioactivity, the yearly dose may be twice this amount. Although an additional yearly dose of 100 mrem from man-made sources would impose a considerable genetic burden on society, this is considered tolerable in view of the benefits, especially since the variation in background over the world is of this order. This additional dose is the maximum permissible annual dose (MPD) to the population as a whole. Contributions to the average yearly gonad dose from artificial sources are given in Table 12.4, showing that the total dose is significantly less than 100 mrem per year.

Table 12.4 Contributions to average annual gonad dose from artificial sources.

Artificial radiation source	Average gonad dose/mrem yr^{-1}
Diagnostic and therapeutic radiology	25
Fall-out	5
Occupational exposure	$0 \cdot 3$
Others (luminous wrist watches, television, etc.)	less than 1
Total	about 30

For people who work in a radiation environment the MPD to the whole body is 5 rem per year, since it is realised that in exceptional (and rare) circumstances work of a more hazardous nature may be required. In this situation the hands alone may receive up to 75 rem per year.

12.6 Protection from external sources

Sources which constitute radiation hazards may be either external or internal. A source on the laboratory bench is an external source relative to the body, but if radioactivity is ingested or inhaled, the contamination is then internal.

Protection against external α- and β-emitters is afforded by the distance between body and source and also by clothes and skin, since charged particles travel a relatively short distance in air and are easily absorbed. High energy β-emitters such as $^{32}_{15}P$ present a problem since their radiation has a long range and may induce cataract formation in the lens of the eye (Table 12.2). Such sources are conveniently manipulated behind a perspex shield sufficiently thick to absorb the radiation.

External photon sources cause most concern. There are three basic criteria for protection. Firstly it may be possible to shield the body from the radiation using a barrier of lead, steel or concrete to reduce the intensity of the radiation and hence the dose rate. The second consideration is the distance between source and body. This situation is shown in Fig. 12.6, the amount of radiation from a localized source which the body intercepts decreasing with increasing separation. It can be shown geometrically that the dose rate D is inversely proportional to the square of the separation d, $D \propto 1/d^2$. If D_1 is the dose rate at separation d_1, and D_2 that at separation d_2 then

$$D_1 d_1^2 = D_2 d_2^2 \tag{12.1}$$

This is a mathematical statement of the inverse-square law of the radiation intensity. A third factor offering protection is the time spent in the vicinity of the source, the occupancy factor. The quicker a manipulation is completed, the less is the radiation risk.

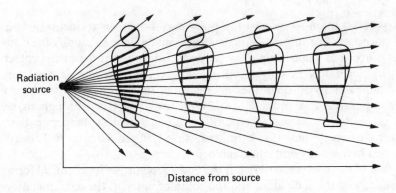

Radiation source

Distance from source

Figure 12.6 Dose reduction with distance (the inverse-square law).

Example 12.1

A γ-source produces a dose rate of 2·0 rem h^{-1} at a distance of 1·0 m. (a) How far away must an occupational worker be so as not to exceed the MPD? (b) If the occupancy is 20 h per week how far away must he work? (c) What thickness of lead ($\mu = 60$ m^{-1}) is required as a shield to enable him to work at 1·0 m for a full 40 h week?

(a) The MPD for an occupational worker is 5·0 rem yr^{-1} = 100 mrem wk^{-1} = 2·5 mrem h^{-1} for 40 h week. Using Equ. 12.1 with $D_1 = 2·0$ rem h^{-1}, $d_1 = 1·0$ m and $D_2 = 2·5 \times 10^{-3}$ rem h^{-1} we have

$$d_2^2 = 2·0 \text{ rem h}^{-1} \times (1·0 \text{ m})^2/(2·5 \times 10^{-3} \text{ rem h}^{-1})$$

$$d_2^2 = 800 \text{ m}^2$$

and

$$d_2 = 28 \text{ m}$$

(b) For 20 h week, MPD = 5·0 mrem h^{-1}. Again, using Equ. 12.1

$$d_2^2 = 2·0 \text{ rem h}^{-1} \times (1·0 \text{ m})^2/(5·0 \times 10^{-3} \text{ rem h}^{-1}) = 400 \text{ m}^2$$

and

$$d_2 = 20 \text{ m}$$

(c) As dose rate is proportional to radiation intensity, it follows the same decrease with distance as Equ. 11.8. When D_1 is the dose rate with no shielding, the dose rate D with a shield L metres thick is $D = D_1 e^{-\mu L}$ or $\log_e(D_1/D) = \mu L$. Putting $D_1 = 2·0$ rem h^{-1}, $D = 2·5 \times 10^{-3}$ rem h^{-1} and $\mu = 60$ m^{-1} we have

$$L = \log_e(800)/60 \text{ m}^{-1}$$

or

$$L = 0·11 \text{ m}$$

12.7   Protection from internal sources

The entry of a radionuclide into the body may be either accidental or for medical purposes. For internal sources, little can be done to modify the dose received, since it is impossible to move away from the source or to protect the body by shielding. The only practicable method of protection is to prevent entry of radioactivity in quantities greater than those considered to be an acceptable hazard. The probability that photons will pass through tissue without interaction makes this type of contamination less dangerous than that where particulate radiation is emitted, since charged particles will always be absorbed in the body and cause biological damage.

The fate of radioactive material in the body depends on its chemical form, the physiology of the individual and the route of entry. If the contamination is inhaled it may subsequently be exhaled, or trapped in the mucous lining of the trachea and swallowed, or deposited in the lungs or dissolved and passed through the alveolar walls into the blood. If radioactive material is eaten it is either absorbed through the intestine wall or excreted. Alternatively, a skin wound might provide a direct route into the blood stream.

After entry, radionuclides such as tritium and $^{22}_{11}Na$ are distributed uniformly throughout the body, but most are concentrated in one particular organ. This is called the critical organ and is the organ whose damage by a particular contamination results in the greatest damage to the body as a whole. Examples are given in Table 12.5. It can be seen that iodine is concentrated in the thyroid, phosphorus and strontium in bone and carbon in fat. The liver is the critical organ for cobalt and zinc whereas the kidneys, in filtering the blood, accumulate any mercury or gold.

Table 12.5   Critical organs for several elements.

Element	Critical organ	Element	Critical organ	Element	Critical organ
H	Whole body	K	Muscle	I	Thyroid
C	Fat	Fe	Spleen	Au	Kidney
Na	Whole body	Co	Liver	Hg	Kidney
P	Bone	Zn	Liver	Po	Bone
S	Testis	Sr	Bone	Ra	Bone

The dose rate received by the critical organ is proportional to the amount of radioactivity in it at any time. At some later time, this dose rate is reduced because such radioactivity decreases exponentially at a rate characterized by its half-life, $T_{1/2}$ (see § 11.3). Also the element concerned may be biologically removed from the organ. This biological excretion is also approximately exponential. The biological half-life, T_B, is the time for half the element to be excreted. These two processes combine to give an effective half-life, T_{eff}, such that

$$1/T_{eff} = 1/T_{1/2} + 1/T_B \qquad (12.2)$$

This effective half-life is the time required for one-half the initial activity to be removed from the organ by a combination of physical decay and biological excretion.

Example 12.2

If the radioactive half-life of $^{131}_{53}I$ is 8·0 days and its biological half-life is 138 days, determine the effective half-life. Calculate the elapsed time before the dose rate is reduced to one-fifth its initial value.

Using Equ. 12.2 with $T_{1/2} = 8\cdot0$ d and $T_B = 138$ d, $1/T_{eff} = 1/8\cdot0$ d $+ 1/138$ d $= (0\cdot125 + 0\cdot007)$ d$^{-1} = 0\cdot132$ d^{-1}. Therefore

$$T_{eff} = 1/(0\cdot132 \text{ d}^{-1}) = 7\cdot6 \text{ days}.$$

The dose rate is proportional to the activity in the body and follows the same exponential decrease with time as Equ. 11.7. Hence $D = D_0 e^{-\lambda t}$ where $\lambda = \lambda_{eff} = \log_e(2)/T_{eff} = (0.693/7.6 \text{ d})$. Since $\log_e(D_0/D) = \lambda t$, putting $D_0/D = 5$ and re-arranging gives

$$t = \log_e(5)/(0.693/7.6 \text{ d})$$

or

$$t = 17.6 \text{ days}$$

The variation of dose rate to the critical organ with time following an intake of radioactive material is shown in Fig. 12.7. The initial rapid rise in the curve occurs during the time the contamination is being transferred to the critical organ. Most of the material has reached the organ at the peak of the curve and the dose rate is a maximum here (D_0). Subsequently the dose rate D decreases exponentially with time t. The dose to the organ between times t_1 and t_2 is equal to the area under the curve between these times, and the total dose to which a given intake of a radionuclide will commit the critical organ (the dose commitment) is obtained from the total area under the curve for all time.

Radionuclides are classified according to their toxicity, which depends on the type of radiation emitted, the critical organ and the effective half-life. The very highly toxic group includes $^{226}_{88}\text{Ra}$ and $^{233}_{92}\text{U}$ which are deposited in the skeleton and emit α-particles. These have a high RBE for bone cancer induction. Biological exchange from the bone is extremely slow and since the physical half-life of such isotopes is several thousand years, the effective half-life is greater than the lifetime of the body. The dose rate remains constant at its maximum value for all time, as illustrated in Fig. 12.7. The lowest toxicity group contains tritium, $^{14}_{6}\text{C}$ and $^{18}_{9}\text{F}$, which are low energy β-emitters with a high biological turnover.

One very highly toxic nuclide is $^{90}_{38}\text{Sr}$ which, although emitting the relatively less harmful β-particles, has a half-life of 28 years and is deposited in bone and teeth. This isotope of strontium is a fission product and occurs in fall-out from nuclear explosions where it contaminates the surfaces of plants directly

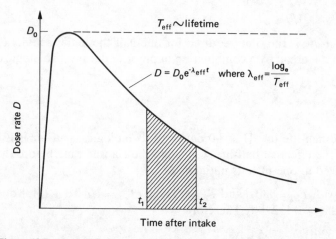

Figure 12.7 Accumulation and excretion of a radionuclide by the critical organ. The dose to the critical organ between times t_1 and t_2 is equal to the area shaded.

or enters the soil to be taken up by the plants. Cows eating grass may accumulate strontium which is then concentrated in the milk to be ingested by
man, or contaminated vegetables may be eaten directly as part of the human
diet. Since strontium is chemically similar to calcium, it follows a similar
metabolic path in the body and is desposited in the bone structure and teeth.
A very simple food chain for $^{90}_{38}$Sr is shown in Fig. 12.8.

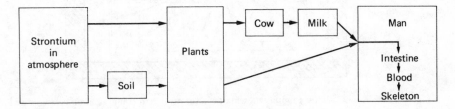

Figure 12.8   Stages in human intake of strontium from fall-out.

The maximum permissible annual intake (MPAI) of any radionuclide is that
intake which results in an annual dose commitment to the critical organ
equal to its annual MPD. Since the MPAI depends on the route of entry into
the body separate values are quoted for inhalation and ingestion. For example,
the MPAI for ingested tritium, distributed uniformly in the body is $2.6 \times 10^4 \,\mu$Ci.
If an occupational worker is estimated to have ingested $1.0 \times 10^4 \,\mu$Ci of tritium,
this is two-fifths of the MPAI and results in two-fifths of the annual MPD for the
whole body. This contribution is therefore 2 rem and the worker must be
limited to a further dose of 3 rem to remain within the MPD of 5 rem per year.

Example 12.3

On average a person inhales 10 m³ of air per working day. If the MPAI for
inhaled tritium is $1.2 \times 10^4 \,\mu$Ci, calculate the maximum permissible concentration in air, (MPC)$_a$, of tritium such that the MPAI is not exceeded.

(MPC)$_a$ = MPAI/volume breathed per year. Assuming a five day working week,
50 weeks per year

$$(MPC)_a = \frac{1.2 \times 10^4 \,\mu\text{Ci yr}^{-1}}{50 \text{ wk yr}^{-1} \times 5 \text{ d wk}^{-1} \times 10 \text{ m}^3 \text{ d}^{-1}}$$

or

$$(MPC)_a = 4.8 \,\mu\text{Ci m}^{-3}$$

Should accidental internal contamination occur, the amount of intake is
usually unknown and must be measured before the dose received can be estimated. The γ-activity in an individual may be measured with a whole-body
counter. This consists of an array of scintillation detectors (see § 11.6) positioned around the body as shown in Fig. 12.9. The contamination is estimated
from the total count rate recorded. For α- or β-contamination, where the radia-

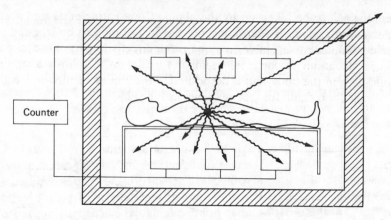

Figure 12.9 A whole-body counter, showing detector array and contamination concentrated in the critical organ. The lead shield reduces environmental background.

tion cannot be detected externally, samples of urine and faeces may be monitored for activity and the contamination estimated from knowledge of the fate of that particular element in the body.

Problems

12.1 High dose irradiation of a polymer solution may induce gel formation, as represented by a decrease in solubility. The data below are taken from measurements on a polystyrene solution. Plot a graph of log (dose) against log (% solubility) and show that the data at these doses are represented by the equation: % solubility $= AD^n$, where n is the gradient of the line and A is constant. Find the value of n from the graph and show that the units of A are rad^{-1}. What is the form of the curve below 1.25×10^7 rad?

Dose/10^7 rad	1.25	1.50	1.75	2.00	2.25	2.50
% solubility	100	80	70	55	45	40

12.2 The survival rate for a population of mammalian cells irradiated in culture is given below at several doses. Plot the results in suitable form to determine the dose required to kill 90% of the cells.

Dose/rem	30	60	90	120	150	180	210	240
% surviving	71	50	35.5	25	18	12.5	9	6.5

12.3 Radiation can kill ordinary, healthy cells. How is this problem overcome in radiotherapy of malignant tumours?

12.4 The dose rate from a γ-source is 2.0 mrem per minute at a distance of 0.50 m. How far away must an occupational worker stand to receive the maximum permissible dose, based on a 40 h working week? Estimate the thickness of steel required to allow him to stand $1 \cdot 0$ m away from the source if the linear absorption coefficient of steel for the radiation is 60 m^{-1}.

12.5 A quantity of $^{203}_{80}$Hg is swallowed and distributed in the body. If the physical half-life of this nuclide is 46 d and the biological half-life is 10 d, find the effective half-life. What time elapses before the activity in the body is reduced to 1/32 the initial activity?

12.6 The physical half-life of $^{32}_{15}$P is $14 \cdot 3$ d and its biological half-life is 1155 d. What is its effective half-life? The maximum permissible concentration of this nuclide in water is $5 \cdot 0 \times 10^{-4} \mu$Ci ml^{-1} and a person drinks $1 \cdot 1$ litres of water at this concentration each day. If 75% of the activity ingested passes into the blood stream and 50% of this is deposited in bone, calculate the activity deposited in the bone each day.

12.7 The effective half-life of tritium is 12 d. If an accidental inhalation of tritium produces an initial dose rate of $1 \cdot 0$ rem per day to the body, plot a curve of the variation of dose rate with time over the 20 days after inhalation and estimate the total dose received during this period from the area under the curve.

12.8 Give reasons for the suitability of a whole-body counter for estimating internal contamination from $^{22}_{11}$Na, which emits γ-rays of energy $1 \cdot 28$ MeV and has a half-life of $2 \cdot 6$ years.

Appendix A: Quantities, Units and Conversion Factors

The International System of Units (SI) is based on the metre (m), the kilogram (kg), the second (s), the ampere (A), the kelvin (K) and the candela (cd). The last of these is the unit of luminous intensity. Some quantities are expressed in terms of the base units and some in terms of derived units. The factors used for converting commonly used quantities into other systems units, the derived units and the multiples of units that are often used are given in the tables below.

Quantity	SI unit	Conversion factor*			
Length	m	1×10^2 cm	3·28 ft	39·37 in	6·2 × 10^{-4} mile
Mass	kg	1×10^3 g	2·21 lb	35·27 oz	6·02 × 10^{26} u
Time	s	1·67 × 10^{-2} min	2·78 × 10^{-4} h	1·16 × 10^{-5} day	3·17 × 10^{-8} yr
Angle	radian	57·3 degree			
Area	m^2	1×10^4 cm^2	10·76 ft^2	1·55 × 10^3 in^2	
Volume	m^3	1×10^6 cm^3	35·31 ft^3	6·1 × 10^4 in^3	1 × 10^3 litre
Density	kg m^{-3}	1 × 10^{-3} g/cm^3	6·24 × 10^{-2} lb/ft^3	3·61 × 10^{-5} lb/in^3	
Speed	m s^{-1}	1×10^2 cm/s	3·28 ft/s	2·24 mile/h	3·6 km/h
Force	kg m s^{-2}	1×10^5 dyn	2·25 × 10^{-1} lbf		
Pressure, stress	kg $m^{-1}s^{-2}$	10 dyn/cm^2	1·45 × 10^{-4} lbf/in^2	7·5 × 10^{-3} mm Hg	9·87 × 10^{-6} atm
Energy, work	kg $m^2 s^{-2}$	1×10^7 erg	7·38 × 10^{-1} ft lbf	2·39 × 10^{-1} cal	6·24 × 10^{18} eV
Power	kg $m^2 s^{-3}$	1×10^7 erg/s	7·38 × 10^{-1} ft lbf/s	2·39 × 10^{-1} cal/s	1·34 × 10^{-3} hp

*A more comprehensive selection of conversion factors can be found in *Handy Matrices of Unit Conversion Factors for Biology and Mechanics* by C. J. Pennycuick (Edward Arnold), 1974.

Quantity	SI Unit	Derived unit and symbol
Force	$kg\ m\ s^{-2}$	newton (N)
Pressure, stress	$kg\ m^{-1}\ s^{-2}$	$N\ m^{-2}$ or pascal (Pa)
Energy, work	$kg\ m^2\ s^{-2}$	joule (J)
Power	$kg\ m^2\ s^{-3}$	watt (W)
Frequency	s^{-1}	hertz (Hz)
Electric charge	$A\ s$	coulomb (C)
Potential difference	$kg\ m^2\ s^{-3}\ A^{-1}$	volt (V)
Electric field intensity	$kg\ m\ s^{-3}\ A^{-1}$	$V\ m^{-1}$ or $N\ C^{-1}$
Magnetic flux density	$kg\ s^{-2}\ A^{-1}$	tesla (T)
Electrical resistance	$kg\ m^2\ s^{-3}\ A^{-2}$	ohm (Ω)
Electrical capacitance	$kg^{-1}\ m^{-2}\ s^2\ A^2$	farad (F)
Electrical inductance	$kg\ m^2\ s^{-2}\ A^{-2}$	henry (H)
Electrical conductance	$kg^{-1}\ m^{-2}\ s^3\ A^2$	siemens (S)

Prefix	Symbol	Multiplying factor
giga	G	10^9
mega	M	10^6
kilo	k	10^3
centi	c	10^{-2}
milli	m	10^{-3}
micro	μ	10^{-6}
nano	n	10^{-9}
pico	p	10^{-12}

Appendix B: Scalar and Vector Quantities

Many quantities have direction as well as magnitude, they are called vectors. Scalar quantities have magnitude but no direction. Some vector and scalar quantities are shown below.

Vectors	Scalars
Displacement	Mass
Velocity	Speed
Acceleration	Time
Momentum	Frequency
Force	Energy, work
Stress	Charge
Pressure	Current
Deformation	Power

Numerical Answers to Problems

CHAPTER 1 (4) $6 \cdot 7$ N m^{-2}, $6 \cdot 7$ Pa; (5) 60 m; (6) No; (7) 20 m s^{-1}, 99 m s^{-2}, $6 \cdot 9 \times 10^3$ N; (8) 22 m s^{-1}, $1 \cdot 6 \times 10^3$ N; (9) $7 \cdot 0 \times 10^2$ rad s^{-1}, $6 \cdot 7 \times 10^3$ rev/min; (12) $F_y = (2 \cdot 8 \pm 0 \cdot 1) \times 10^2$ N, $F_x = (0 \cdot 0 \pm 0 \cdot 1) \times 10^2$ N; (13) $1 \cdot 5$ kW, $2 \cdot 0$ hp; (14) 260 J min^{-1}, $5 \cdot 8 \times 10^{-3}$ hp; (15) approximately 1 m, $4 \cdot 4$ m s^{-1}.

CHAPTER 2 (1) $1 \cdot 0$ Hz, $1 \cdot 0$ s; (2) $9 \cdot 9$ N; (4) $5 \cdot 0 \times 10^{-3}$ m, $3 \cdot 3 \times 10^{-2}$ s, 30 Hz, $1 \cdot 8 \times 10^2$ m s^{-2}, 18 g's; (5) $7 \cdot 7 \times 10^3$ N m^{-1}, 51 mm; (6) $0 \cdot 11$; (7) $0 \cdot 113 \pm 0 \cdot 003$; (10) $0 \cdot 63$ N m^{-2}, $0 \cdot 89$ N m^{-2}; (11) $4 \cdot 7$ s, $5 \cdot 4$ μs; (12) $5 \cdot 8$ ms, $1 \cdot 2$ ms; (13) 83 dB.

CHAPTER 3 (1) $3 \cdot 8 \times 10^{-19}$ J, $1 \cdot 33 \times 10^{-28}$ J; (3) $1 \cdot 1 \times 10^{-34}$ kg m^2 s^{-1}, $2 \cdot 0 \times 10^{-24}$ kg m s^{-1}, $2 \cdot 2 \times 10^6$ m s^{-1}; (4) $0 \cdot 54$ eV, $1 \cdot 3 \times 10^{14}$ Hz; (5) $9 \cdot 5 \times 10^{-37}$ m, $1 \cdot 1 \times 10^{-25}$ m, $3 \cdot 2 \times 10^{-10}$ m, $3 \cdot 7 \times 10^{-63}$ m; (7) $0:25 \cdot 6\%$, C:$9 \cdot 5\%$, H:63%, N:$1 \cdot 3\%$, Ca:$0 \cdot 2\%$, P:$0 \cdot 2\%$, S:$0 \cdot 05\%$, K:$0 \cdot 03\%$, Na:$0 \cdot 04\%$, Cl:$0 \cdot 02\%$, Mg:$0 \cdot 01\%$, Fe:$6 \times 10^{-4}\%$; (8) $5 \cdot 1$ eV, $9 \cdot 8$ eV, $1 \cdot 25 \times 10^{14}$ Hz; (10) 2.99×10^{-26} kg, $1 \cdot 53 \times 10^{-25}$ kg; (11) $5 \cdot 0 \times 10^4$ m s^{-1}, 5.2 mm; (12) 27.994 915 u, 28.006 148 u, 28.031 300 u.

CHAPTER 4 (1) $8 \cdot 3 \times 10^{-2}$ eV; (2) $E_{650}/E_{500} = 0 \cdot 87$; (3) 280 units; (4) 400 ppm; (5) $2 \cdot 3 \times 10^{-26}$ kg, 1.3×10^{-45} kg m^2, $1 \cdot 3 \times 10^{10}$ Hz; (6) $6 \cdot 8 \times 10^{-23}$ J, 20×10^{-23} J, 41×10^{-23} J, 68×10^{-23} J; (7) 1850 N m^{-1}, $0 \cdot 265$ eV; (8) $3 \cdot 0 \times 10^{-20}$ J, $1 \cdot 1 \times 10^{-11}$ m; (9) $0 \cdot 25$, $0 \cdot 53$, $0 \cdot 77$, $0 \cdot 89$, $1 \cdot 14$ ppm; (11) $8 \cdot 1$ kV, $1 \cdot 35 \times 10^4$ m^{-1}.

CHAPTER 5 (2) $1 \cdot 0 \times 10^8 \Omega$, $4 \cdot 1 \times 10^6 \Omega$, $1 \cdot 0 \times 10^6 \Omega$, $2 \cdot 6 \times 10^5 \Omega$; (3) 41Ω, $1 \cdot 6 \Omega$, $0 \cdot 41 \Omega$, $0 \cdot 10 \Omega$; (4) $I = 8$ A$_{rms}$; (5) $0 \cdot 42$ m; (7) $0 \cdot 0016$ μF, $2 \cdot 2$; (8) $1 \cdot 85$ s, $8 \cdot 6 \times 10^{-2}$ Hz; (11) $0 \cdot 013$ μF, $0 \cdot 026$ μF mm^{-2}.

CHAPTER 6 (1) 275 Hz; (2) 10 MΩ; (4) $0 \cdot 48$ Ω; (8) $(1 \cdot 1 \pm 0 \cdot 1) \times 10^{-2}$ N; (9) $0 \cdot 60$ mV, 120 mV; (10) $1 \cdot 0$ V$_{rms}$, $0 \cdot 13$ V$_{rms}$, $V_S/V_N = 7 \cdot 7$.

CHAPTER 7 (1) Parallel to incident beam, displaced laterally by $10 \cdot 4$ mm; (5) $22 \cdot 3$ mm, $22 \cdot 3$ mm; (6) 64 dioptres, $22 \cdot 3$ mm; (7) $0 \cdot 027$; (9) 31 μm; (10) 50 m; (13) $-6 \cdot 7 \times 10^{-2}$ rad kg^{-1} m^2, 30 kg m^{-3}.

CHAPTER 8 (1) $1 \cdot 0$ mm, $2 \cdot 5$ mm; (6) $0 \cdot 31$ μm, $0 \cdot 26$ μm, $0 \cdot 15$ μm, $0 \cdot 61$ nm; (7) S_R, $2 \cdot 4$ μm, $0 \cdot 44$ μm, $0 \cdot 24$ μm; S_R'', $9 \cdot 6 \times 10^1$ μm, $1 \cdot 7 \times 10^2$ μm, $2 \cdot 3 \times 10^2$ μm; (8) $I_{1 \cdot 4}/I_{0 \cdot 7} = 4$; (11) $0 \cdot 37$ λ.

CHAPTER 9 (5) $0 \cdot 0$ °C, $3 \cdot 3 \times 10^3$ J kg^{-1} K^{-1}; (6) $3 \cdot 6$ °C; (7) 320 W m^{-2}; (8) 44 W, $0 \cdot 67$ °C/h; (9) 22 W, $4 \cdot 8$ W, 94 W, 140 W; 8%, 2%, 36%, 54%; (12) 29×10^{-3} N m^{-1}; (13) $1 \cdot 83 \times 10^{-2}$ N, $1 \cdot 87 \times 10^{-3}$ kg; (14) 13 m.

CHAPTER 10 (2) $1 \cdot 8 \times 10^3$, $4 \cdot 0 \times 10^6$ N m^{-2}; (4) (a) $1 \cdot 5 \times 10^8$ N m^{-2}, (b) $3 \cdot 9 \times 10^7$ N m^{-2}; (5) it breaks; (6) $1 \cdot 2 \times 10^7$, $6 \cdot 7 \times 10^7$, $2 \cdot 4 \times 10^7$, $1 \cdot 5 \times 10^7$ N m^{-5} s; $1 \cdot 2 \times 10^8$ N m^{-5} s; (7) $3 \cdot 8 \times 10^4$ N m^{-2}; (8) $2 \cdot 8$ m s^{-1}; (10) $\eta = 0 \cdot 14(\Delta \gamma / \Delta t)^{-0 \cdot 66}$; (12) 5×10^4 N m^{-2}, 5×10^3 N m^{-2}, $3 \cdot 5 \times 10^3$ s.

CHAPTER 11 (1) $63 \cdot 6$; (3) $9 \cdot 3 \times 10^6$ dps; (4) $1 \cdot 3$ mCi, $4 \cdot 8 \times 10^7$ Bq, 174 days; (5) $0 \cdot 069$ min^{-1}, 10 min; (6) $2 \cdot 37 \times 10^{-10}$ min^{-1}, $1 \cdot 61 \times 10^4$ dpm; (7) approximately 3900 years; (8) 25%; (9) $2 \cdot 3 \times 10^5$, $4 \cdot 3 \times 10^3$, $-3 \cdot 7 \times 10^{-14}$ C, $-6 \cdot 9 \times 10^{-16}$ C; (10) 9×10^{16} J, $5 \cdot 6 \times 10^{35}$ eV; (12) 101%, $1 \cdot 3$%; (13) $2 \cdot 5$, $4 \cdot 5$, $0 \cdot 5$, $9 \cdot 0$ μCi; (14) 50 kg; (15) $0 \cdot 13$ Gy, $2 \cdot 0$ Gy, 15, 200 rem.

CHAPTER 12 (1) $-1 \cdot 3$; (2) 200 rem; (4) $3 \cdot 5$ m, 42 mm; (5) $8 \cdot 2$ days, 41 days; (6) $14 \cdot 1$ days, $0 \cdot 2$ μCi; (7) 12 rem.

Index